Quaking

*Art, science and the community:
a collaborative approach to water pollution*

Penny Kemp & John Griffiths

JON CARPENTER

Our books may be ordered from bookshops or (post free) from
Jon Carpenter Publishing, 2 The Spendlove Centre, Charlbury,
England OX7 3PQ

Please send for our free catalogue

Credit card orders should be phoned or faxed to 01689 870437
or 01608 811969

Our US distributor is Paul and Company, PO Box 442, Concord, MA 01742
(phone 978 369 3049, fax 978 369 2385)

First published in 1999 by
Jon Carpenter Publishing
2, The Spendlove Centre, Charlbury, Oxfordshire OX7 3PQ
☎ 01608 811969

ISBN 1 897766 57 2

Printed and bound in England by J. W. Arrowsmith Ltd., Bristol
Cover and colour pages printed by KMS Litho, Hook Norton

Quaking Houses

Contents

Foreword   *Professor Graham Ashworth*   7

A personal introduction to Seen & Unseen   *Lucy Milton*   8

About this book   9

Acknowledgements   10

1   Contexts   13
2   The Sunderland scarf   29

Colour plates
Quaking Houses: a pictorial record   33

3   Genesis of the project   49
4   Gavinswelly and beyond   57
5   Visions of utopia   63
6   Completing the project: the artist's perspective   77
7   The project as an actuality   95
8   The scientific and engineering context of the Quaking Houses
        community wetland   *Dr P. L. Younger*   109
9   Art in Seen & Unseen: Context and evaluation   *Malcolm Miles*   119
10 Lessons and conclusions   133

Appendix: Scientific data from Quaking Houses wetland   139

Foreword

Professor Graham Ashworth
Chairman & Chief Executive, Going for Green

'Think global, act local' has become such a mantra for the environmental movement that it is sometimes possible to lose sight of what this actually means in practice. The Quaking Houses wetland, however, is an excellent example of what can happen when people do act locally with the broader good also in mind. The project sprang from a community's concern for its local environment, but a concern always informed by an awareness of the wider issues of the environment in Quaking Houses itself, County Durham, and beyond. Their efforts, complemented by the innovative work carried out at the University of Newcastle upon Tyne and the imaginative perspective of the artist-in-residence Helen Smith, have brought about the completion of a project that enhances the local landscape, has a valuable educational role, and has effectively cleaned up the polluted Stanley Burn—with implications for the cleanliness of the River Wear into which it flows, and, ultimately, the North Sea. The Quaking Houses wetland project demonstrates that however intractable a problem may seem, people, by working together with imagination and dedication, can make a local difference and in the process offer encouragement and education to a wider audience.

A personal introduction to Seen & Unseen

Lucy Milton, Artistic Director, Artists' Agency

It all started in my allotment in the early '90s, when I noticed that my vegetables would only thrive if I planted them much earlier in the year than usual. I began to worry about global warming and its effect on future generations; in particular, and selfishly, that my daughter would inherit a world incapable of sustaining itself. I besieged everyone I knew with questions, got even more worried, and began collecting a growing number of newspaper articles. At first I felt hopeless—there was nothing I could do as an individual to make a difference. Then I thought that if I could use my knowledge of the arts to develop a local project tackling a major pollution problem in the Northern Region, this might contribute to a critical mass of positive solutions. These solutions could connect the local to the global and encourage us to realise that if we work together we can all do our small bit to help us preserve our world for future generations. Thus Seen & Unseen was born.

I felt that the most effective way to tackle environmental problems was for people with different areas of expertise—artists, scientists and communities— to pool that knowledge and explore new ways of working together. I chose water pollution both because of its importance as an issue, but also because of water's potential for metaphor. I obtained funding from Northern Arts for a feasibility study, met Paul Younger, and he introduced me to members of the Quaking Houses Environmental Trust. We decided to work together and share information about what happened—the problems as well as the successes to help others on their way. Hence this book.

About this book

This book is an attempt to describe an unprecedented and, in terms of the relationships between the participants, very complicated enterprise. This is reflected in the layout of the book. The early chapters, by John Griffiths, set the scene for the Quaking Houses project, describing the village and its mining legacy, the nature of the pollution problem in the Stanley Burn, and the construction of both pilot and full-scale wetlands. The subsequent chapters, by Penny Kemp, give a blow-by-blow account of how the project moved forward and illustrate the highly personal and often argumentative workings of the community-scientist-artist collaborative process at 'the sharp end', highlighting the benefits and pitfalls of this way of working. Dr. Paul Younger, of the University of Newcastle upon Tyne, describes developments in passive treatment of minewater pollution and the importance of the Quaking Houses wetland in this developing technology, while Malcolm Miles of Oxford Brookes University assesses the project's place in the context of the broad field of environmental art.

Penny Kemp
John Griffiths

Acknowledgements

In memory of the late Terry Jeffrey, who gave his all to Quaking Houses.

With huge thanks to all the people who made this project possible, especially Terry Jeffrey, Chas Brooks, Jozefa Rogocki, Paul Younger, Adam Jarvis, Janet Ross, David Butler, all at PLATFORM and Helen Smith.

And thanks to everyone else who contributed to this project and to this publication in so many different ways, including:

Almudena Ordoñez Alonso, Chris Bailey, Colin Basey, Julia Bell, Tommy Cole, Walter Cooper, John Cram, Lee Dalby, the local studies section at Durham Library, Douglas Glendinning, Dan Grierson, John Griffiths, Neville Hart, Ian Jeffrey, Penny Kemp, John Knapton, Ralph Little, all at Loftus Development Trust, Alan McCrea, the late Jamie McCullough, Michael McManus, Malcolm Miles, Caroline Mitchell, Kath Nichols, Paul Nugent, Lynne Otter, Mark Pearson, members of Quaking Houses Village Hall Association, Mike Rafter, Diane Richardson, Robert Shiel, Steve Smith, Joy Taylor, all at the Virtual Reality Centre, University of Teesside, Martin Weston, Colin Wilkinson, Gail Wilson, Maureen Wilson, Peter Woodward, and of course, all the project's funders. Thanks are also due to Julie Ross, whose written assessment of the project, produced by courtesy of the Glasgow School of Art, formed a starting point and valuable source of information for much of this publication.

Lucy Milton, Artists' Agency

The authors wish to express their gratitude to the North of England Open Air Museum, Beamish, for permission to reproduce photographs of Quaking Houses and its surrounding mines; to Durham County Record Office for permission to reproduce Ordnance Survey map extracts; and to William Heinemann for permission to reproduce the extracts from *English Journey* by J. B. Priestley.

The wetland project forms part of the Seen & Unseen project which has been developed by Artists' Agency in collaboration with Quaking Houses Environmental Trust and the University of Newcastle upon Tyne Civil Engineering Department. The main funders of Seen & Unseen are Northumbrian Water's Kick-Start Scheme, which was an award winner under The Pairing Scheme in its support of Seen & Unseen (The Pairing Scheme is a government scheme managed by Arts & Business), the Department of the Environment, Transport and the Regions' Environmental Action Fund (linked to its "Are you doing your bit?" environmental awareness campaign) and the Virtual Reality Centre at the University of Teesside linked to ERDF funding—RESIDER. Other major funders have been Northern Arts, Arts Council of England, Shell Better Britain, the Rural Development Commission, the Lankelly Foundation, Thompson's of Prudhoe Environmental Trust and Northumbrian Water Environmental Trust. Valuable support has also been received from AN, the University of Northumbria at Newcastle, NUWATER Consulting Services Ltd, the Environment Agency, the University of Sunderland, Derwentside District Council, Quaking Houses Environmental Trust, Durham County Waste Management Company Ltd, Gulbenkian Foundation, Greggs, English Nature, BT, The Hadrian Trust, The Sir James Knott Trust, The Sir John Priestman Charity Trust, CETPASE, Loftus Development Trust, Village Arts, The Hancock Museum, Glasgow School of Art, County Durham Training and Enterprise Council, ARCO, National Power, and British Coal Property.

Artists' Agency, a registered charity (no 700956), is revenue funded by Northern Arts.

In autumn 1999 Artists' Agency changed its name to Helix Arts.

1 *Contexts*

Six hundred and ninety feet above sea level, on a windy ridge a mile south of the former mining town of Stanley, County Durham, and seven and a half miles north-west of the city of Durham with its World Heritage Site Cathedral and Castle, lies Quaking Houses, known to its inhabitants as 'Quakies', a mining village that has outlived the industry that gave it birth, whose distinctive name might arouse momentary recognition but which is unfamiliar to most people even in its home county.

County Durham is not short of former mining villages, but Quaking Houses, so easily overlooked, with its single access road emphatically a place that one has to go to rather than ever pass through, is different. The villages of Durham have known bitter suffering and neglect, and partly as a consequence have developed strong community bonds; in Quaking Houses, these bonds gave rise to a campaigning spirit and determination to improve the local environment which in turn led to the development in the village of a unique project where the community, scientists and artists came together to regenerate the local waterway, the Stanley Burn, by developing a wetland area to cleanse the stream of pollution.

This book tells the story of this project, how it came about, how it worked, what were the benefits and pitfalls, and how the various, very different groups and personalities involved were able to work together for compatible and mutually beneficial ends. To understand this it is necessary to step backwards from the project and look at the village and its surroundings, examining why Quaking Houses exists, how it grew up where it did, and how broader economic and social influences have created the conditions which gave rise to the problems and to the solutions which will be described in the following chapters.

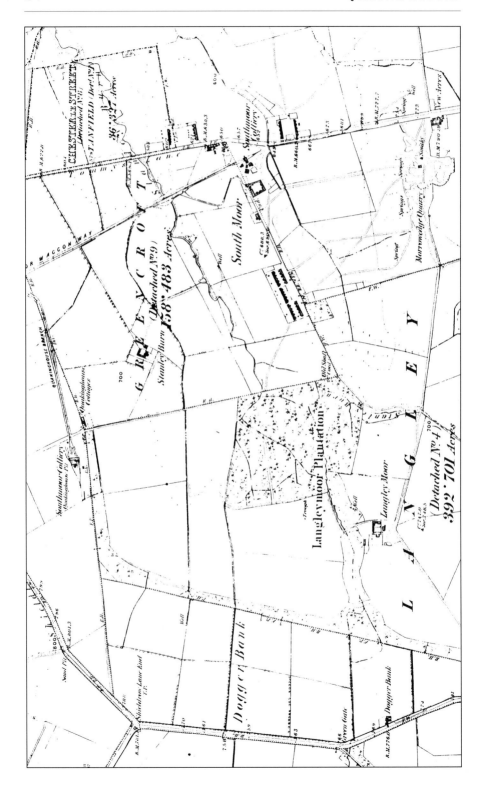

Black diamonds

Coal and County Durham have together a history that stretches back to Roman times: coal ash and unburned fragments of coal have been found in the remains of Roman sites in the region. Recorded coal mining dates from the middle ages, with references to coal workings in the thirteenth century; mining is known to have taken place in the Derwentside area since around 1350, and the industry was general throughout the County Palatinate of Durham from the fifteenth century. The mines provided the 'black diamonds' that made the mineowners and landowners of the region immensely rich; the grand houses such as Lambton Castle, Lumley Castle, Wynyard Hall, the great palace of the Marquesses of Londonderry, and even Durham Castle itself, home of the Prince Bishops, could fairly be said to have been carved out of coal. Today, the industry which powered Britain's drive to economic supremacy in the eighteenth and nineteenth centuries is just a wraith. The pits have—one by one—fallen silent; the spoil heaps have been levelled and grassed over or planted with trees; the surface workings demolished. The villages remain—blocks of terraced streets that look as if they belonged to some great industrial city, ripped up by a giant's hand and transplanted to what now appear to be the unlikeliest of rural settings. Quaking Houses is one such village; and apart from its continued existence as a community, by the 1980s Quaking Houses had one further visible legacy from the coal industry: the poisoning of its local stream, the Stanley Burn, by pollutants from a nearby spoil heap.

The names of the industrial and mining villages of County Durham form a sort of bleak poetry: Pity Me, Billy Row, Perkinsville, Inkerman, Fir Tree, Metal Bridge, No Place, Quaking Houses. Quaking Houses is an ominous name; one which, with its suggestions of subsidence and subterranean movement, is entirely suitable for a mining region. It is also obscure in origin. Old maps show a Quaking House, close to where the Annfield Plain by-pass now runs south of New Kyo, and it may be that the village took its name from this. The area was originally known as Old South Moor, and the oldest settlement in the area, built after the William Pit was sunk to the east of the village site in 1839, was known as South Moor Cottages. Half-a-mile to the

Opposite: Extract from Ordnance Survey first edition 6 inches to the mile map showing the Quaking Houses area, as surveyed in 1856–7.
REPRODUCED BY KIND PERMISSION OF DURHAM COUNTY RECORD OFFICE.

east of this, the Quaking House Pit—originally known as the New Shield Row Pit, and subsequently as the Charley Pit—was sunk in 1845. The late Fred Wade, a noted local historian who lived in Third Street, Quaking Houses, also records a shaft in this area dating from 1818, possibly the disused coal shaft shown on the 1857 Ordnance Survey map at the south-eastern corner of Langley Moor Plantation. Near to the Charley Pit this was a short row called Quaking House Cottages, a farm called Quaking House Farm, and the land was known as Quaking Hill. Yet none of this properly explains the name. One theory is that underground working caused build-ings topside to vibrate—but even if true, this cannot have been so unusual a phenomenon in mining areas like this part of Durham. Another is that the name derives from a Quaker who once lived in the block of colliery cottages known as The Barracks, near the east of the village. Some sources refer to 'Quaker House Pit', and Wade gave it as his opinion "that Quaking is derived from Quaker and that at one time a Quaker meeting house was situated near the old pit, and the name Quaking Houses was given to South Moor Cottages to identify them from the other three groups of cottages that then formed the village of South Moor".

Be that as it may, the village was firmly established by the time of the first edition of the 1:2500 scale (25" to the mile) Ordnance Survey maps, surveyed in 1857, albeit unnamed on the maps. The maps show at the east end of the village the four-sided block of cottages known as The Barracks; 'Southmoor Colliery', and four terraced rows parallel with or running off what is now South Moor Road but was then Tommy's Lane, or Tommy's Lonnen, named after Thomas Daglish who had the right to gather horse manure thereabouts. Further west, beyond Bleak House, the colliery manager's residence, was Quaking Houses 'proper', five rows of ten cottages each, three facing northwards and two southwards. Looking at the map, the continuity of layout between these parallel rows and the present-day village is unmistakeable.

Growth of the village was sparked by the sinking of the William Pit in 1839, but others followed: the Charley Pit, as mentioned, in 1845; the Hedley Pit in 1885. Successive editions of the 1:2500 map show how further

Opposite: Extract from Ordnance Survey second edition 6 inches to the mile map showing the Quaking Houses area, as revised in 1895.
REPRODUCED BY KIND PERMISSION OF DURHAM COUNTY RECORD OFFICE.

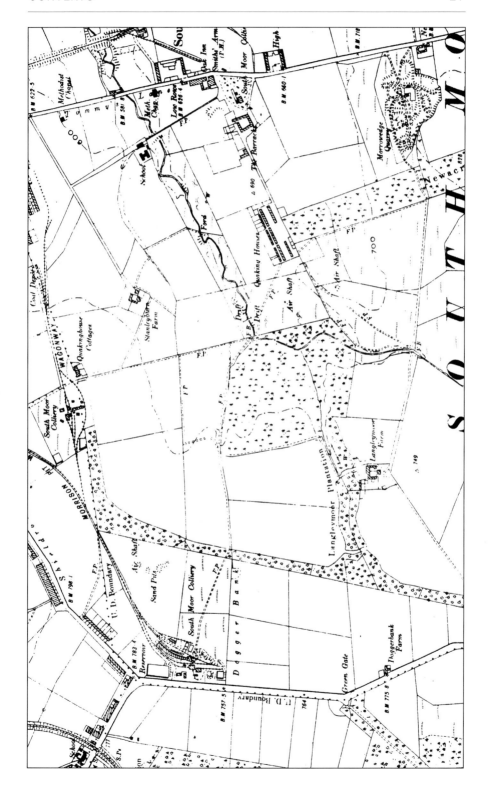

The William Pit at the eastern end of Quaking Houses, sunk 1839, closed 1960.
REPRODUCED BY KIND PERMISSION OF THE NORTH OF ENGLAND OPEN AIR MUSEUM, BEAMISH

mines were established around the village. The second edition, revised in 1895, shows two drift mine entrances, one on each side of the Stanley Burn, just north-west of the village with a wagonway spanning the burn, a small spoil heap alongside, and ventilation shafts on Langley Moor to the south; these were known as the Shield Row drifts. The 1915 revision, published after the end of the Great War in 1920, shows this mine as being greatly expanded, the North and South Drift entrances marked, a third adit to the south (level with the end of First Street), a fan house to ventilate the mine, and an aerial cableway linking the mine to the Louisa Pit to the north. This enterprise was small beer however compared with the two huge mines which were developed west of the village, on the outskirts of Annfield Plain on a tract of land known as Dogger Bank: the Morrison, the North Pit of which was first opened in 1868 by William Bell & Partners and the South Pit shortly after, and, crucial to the story of the Quaking Houses wetland

Opposite: Extract from Ordnance Survey third edition 6 inches to the mile map showing the Quaking Houses area, as revised in 1915.
REPRODUCED BY KIND PERMISSION OF DURHAM COUNTY RECORD OFFICE

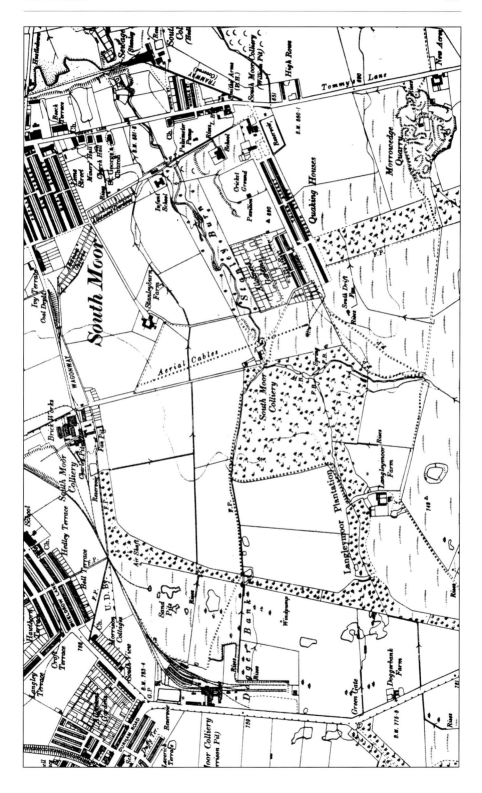

project, the Morrison Busty Pit. Work on the Morrison Busty began in 1922, and the shafts were bottomed out in 1925 on the Brockwell coal seam, 780' below the surface. Coal production began in 1927. These were operated by the South Moor Colliery Co, which after merger with the Hedley Coal Company in 1925 became Holmside and South Moor Collieries Ltd, known as a relatively progressive employer and the only major Durham mining concern not to join the reactionary Durham Coal Owners Association. The Morrison pits were noted not only for coal: witherite—carbonate of barium—was discovered in Morrison North Pit in 1930, one of only a handful of production sites in the world (all in the north-east of England; others included Settlingstones in Northumberland, Brancepeth and Haggs in the Nent Valley). This valuable mineral, extracted from a shaft at Burnhope Road Ends, Maiden Law, was used in paper and linoleum manufacture, paints and printing inks, and a plant was erected at the North Pit to process it, grading the material from clean pieces to ground powder. Morrison Busty, moreover, was the first pit in the country to have a full-time nurse and purpose-built medical centre.

"I thought pitmen lived in the pit!"

"Some years ago a gentleman came north to fill an important position in a large colliery. He arrived late at night, and slept at the inn. Next morning, on looking out he called the landlord… and asked what all those cottages were for. 'Cottages, sir? those are the pitmen's houses.' 'Good gracious!' exclaimed the southerner. 'I thought pitmen lived in the pit!'" [J. R. Boyle, *Comprehensive Guide to the County of Durham* (London, 1892)]

Conditions in mining communities have often seemed utterly mysterious to the outside world. The anecdote cited in Boyle's *Comprehensive Guide* merely confirms the impression received by the author and polemicist William Cobbett in 1832 while on a tour of northern England, that "here is the most surprising thing in the whole world; thousands of men and thousands of horses continually living under ground; children born there, and who sometimes, it is said, seldom see the surface at all, though they live to a considerable age."

The reality, in the early years of the existence of Quaking Houses, would have been bad enough. The tiny cottages of The Barracks were less than fifty yards from the buildings of South Moor Colliery (the William Pit) with its engine, pithead gear, clattering waggonway, spoil heaps and universal filth and dust; the other rows of South Moor scarcely further away. And, as homes, the cottages would have been barely adequate, "little more than brick shelters to keep out the prevailing cold winds and the inclement weather" according to South Moor Local History Group. The cottages were two-roomed: the lower room paved with flagstones, the upper reached by a ladder through a hole in the bare boards of the upper floor; there were no ceilings, maybe a bearable drawback compared with the absence of running water, drainage and sanitary facilities. However, the discovery of greater coal reserves than had initially been expected led the mineowners to invest in better housing, such as two-bedroom cottages, to retain the workforce.

Other facilities developed locally. The area had three pubs—the Mason's Arms (later called the Smith's Arms), and two doors up, the Oak Inn, both shown on the 1857 map (the Mason's Arms was probably considerably older; it had been closed but had reopened as a pub when the mines were sunk), and to the north, the Stag Inn. And, typical of mining communities the length and breadth of the country, religious nonconformity was a strong influence, represented here by the Wesleyan Methodist chapel next to the Stag Inn on Tommy's Lane, opened in 1858. Educational facilities were also provided: Holmside National School opened in 1847 on Wagtail Lane in nearby Craghead, while by 1894 a sizeable elementary school had been built just to the north of the Stanley Burn in South Moor, linked to The Barracks area of Quaking Houses by a footpath along the line of the former South Moor Colliery wagonway.

As the prospects for continued deep mining in the area grew brighter, so improvements were made to the miners' villages. The period between the turn of the century and the 1920s saw Quaking Houses take on much of its present-day appearance, with the reconstruction of the village by the South Moor Colliery Co. The Barracks were demolished, and Third and Fourth Streets built to the west, while the five terraced rows of Quaking Houses 'proper' were extended and rebuilt, as First and Second Streets; to their west, the larger houses of Woodside Terrace and Fellside Terrace were subsequently added for mine officials. The bricks for these improvements were made at the brickworks at the Charley Pit (brickmaking was frequently a companion industry to coal mining, using clay excavated from the pits) and conveyed to the village by an aerial ropeway.

The relative isolation of the village was diminished by motor bus services, and numerous companies such as Hather Bell, Hammell's, Mowbray's, Northern (which still runs there) and the splendidly-named French's Yellow Belly served Quaking Houses; while one measure of relative prosperity is provided by the decision of the Stanley Co-operative Society to open a branch on Third Street in 1921. Even a cricket ground was provided, where Bleak House stood prior to its demolition in 1892; the stones of Bleak House were used for the boundary wall south of the pitch. This was the home from the early twentieth century of South Moor Cricket Club, founded in or around 1884 (the club previously played at New Acres Farm, south of the village), and next to the cricket ground was the football pitch, home of the Quaking Houses 'Lillywhites'. The village even had a sporting world champion—Harry Rostron, the quoits champion.

To the north of the cricket ground and spanning the Stanley Burn, the Memorial Park was opened in 1920 on land donated by the South Moor Colliery Co as a memorial for men killed in the Great War. It came complete with all the amenities of the classic municipal park: putting green, paddling pool, bowling green and bandstand. One of the park's trees provided a more prosaic amenity: a hole in the trunk at a convenient height proved irresistible

Quaking Houses. Third Street under construction.
REPRODUCED BY KIND PERMISSION OF THE NORTH OF ENGLAND OPEN AIR MUSEUM, BEAMISH

for homeward-bound miners to knock out the residues from their carbide lamps. The tree sickened, and eventually had to be felled, to reveal that the corrosive ash had completely hollowed out the inside of the trunk!

This period was the high-water mark of prosperity and confidence; and it was not to last long. Post-war depression, the General Strike of 1926, further strikes in 1928, the Great Depression, shut-downs and wage cuts hit the area—hit all of the coalfields in Britain—very hard. J. B. Priestley visited the Durham coalfield for his *English Journey* of 1934, and what he wrote about East Durham could equally apply to the Quaking Houses district:

> "The first impression of my own that was instantly confirmed was that of the strange isolation of this mining community. Nobody… goes to East Durham. The miner there lives in his own little world and hardly meets anybody coming from outside it… He is isolated geographically. More often than not he lives in a region so unlovely, so completely removed from either natural beauty or anything of grace or dignity contrived by man, that most of us take care never to go near a colliery area. The time he does not spend underground is spent in towns and villages that are monuments of ugliness. I shall be told by some people that this does not matter because miners, never having known anything else, are entirely indifferent and impervious to such ugliness. I believe this view to be as false as it is mean. Miners and their wives and children are not members of some troglodyte race but ordinary human beings and as such are partly at the mercy of their surroundings. I do not want to pretend that they are wincing aesthetes, unnerved by certain shades of green and sent into ecstasies by one particular pink. But the channels of the sense are open to them just as they are to the rest of us. Their environment must either bring them to despair—as I know from my own experience that it frequently does—or in the end it must blunt their senses and taste, harden the feelings and cloud the mind."

The desire of communities for a decent environment to live in, highlighted here by Priestley, is central to the subsequent history of Quaking Houses, but even as he was writing, economic considerations—on a personal level, the problems of day-to-day existence—were of greater significance to the local people than their physical surroundings. Across County Durham, employment in the coal industry had fallen from 165,246 in 1913 to 128,038 by 1930, and

by 1933 was to fall to 98,096. Mines operated at a loss, and in many companies mines were worked on alternate weeks; thus it was proposed that of the
mines operated by the Holmside and South Moor Collieries Co, the Hedley,
William and Louisa pits should work one week; Morrison North and South the
other week. Pre-war rearmament programmes and subsequent wartime
demand brought some respite to the industry, but, for all the optimism
surrounding Vesting Day in 1947, the nationalisation of the mines under the
National Coal Board signalled the start of long-term and inexorable decline.

Changing technologies and the working-out of older pits meant that the
area's economic base began to run down. Morrison South Pit shut in 1945
and the Hedley Pit in 1946. The following year, on August 22 1947, a terrible
accident occurred in the Louisa Pit. This pit worked the middle coal
measures in the area around and beneath Quaking Houses and South Moor
(the upper measures were worked from the Hedley Pit and the lower from
Morrison Busty). A miner had taken illegal smoking material into the pit, and
on lighting a cigarette ignited firedamp (methane gas) which had accumu-
lated. The resulting detonations set off a much larger explosion and
conflagration fuelled by coal dust in the air; of twenty-four men working in
the Hutton seam on that shift, twenty-one died.

During the 1950s coal supplies began to exceed the nation's needs, as
greater reliance was placed on electricity and other energy sources and the
railways switched from steam to diesel or electric traction. Locally this
brought about the final demise of deep coal mining. Morrison North Pit
closed in 1961, the Louisa Pit in 1967 and finally, on 5 October 1973, the last
shift was worked at the Morrison Busty Pit. A centuries-old industry and way
of life had come to an end.

"Derelict, doomed, damned"

The mines had not even all ceased production before officialdom was
writing the death notice for Quaking Houses and dozens of settlements like
it. Durham County Council's 1951 County Development Plan formalised a
policy of encouraging a concentration of the workforce in fewer towns and
villages, supposedly more viable socially and economically than the settlement pattern of small and widely scattered mining villages. All the 357 towns
and villages in the county were allocated to one of four categories: 70 were
category A, where considerable projected expansion in population was held
to merit public investment; 143 were category B, where sufficient investment

The Charley Pit (Quaking Houses Pit), showing the large Waddle fan.
REPRODUCED BY KIND PERMISSION OF THE NORTH OF ENGLAND OPEN AIR MUSEUM, BEAMISH

was recommended to satisfy the needs of a static population; 30 were category C, meriting minimal investment to cope with a slowly declining population; and, notoriously, 114—later increased to 121—were category D, settlements "from which a considerable loss of population may be expected. In these cases it is felt that there should be no further investment of capital on any appreciable scale... This generally means that when the existing houses become uninhabitable they should be replaced elsewhere, and that any expenditure on facilities and services in these communities which would involve public money should be limited to conform to what appears to be the possible future life of existing property in the community." Quaking Houses was such a category D settlement.

Although, in its written analysis of the development plan, the council protested that "there is no proposal to demolish any village, nor is there a policy against village life. It is proposed to remould gradually the pattern of development in the interests of the county as a whole", it is clear that the formal adoption of this policy was a death sentence for Quaking Houses and villages of its type—in the Stanley Urban District, the other category D villages were Craghead, The Middles, Oxhill, West Kyo, East Castle, White-le-Head, Tanfield Lea, Hobson, Lintz and Causey New Row.

Eighteen years on, the slow starvation policy was starting to take effect, documented by journalist John Barr in *New Society* (3 April 1969). Describing the gap-toothed terraces, the boarded-up shops, the muddy unadopted streets in what he called "Durham's murdered villages", Barr typified category D villages as "dejected places with inappropriate names like Eden Pit, Sunniside and Mount Pleasant or with appropriate names like Burnt Houses, The Slack and Pity Me. There the D label has come to mean derelict, doomed, damned." While the 'category D' label was dropped in what would today be described as a piece of 'spin', the policy remained in place (and would continue to remain until 1976, when the first County Structure Plan was drawn up).

In his 1976 book *Durham Villages*, Jarrow-born Harry Thompson described Quaking Houses as "the archetypical pit village that is dying": "Nowadays the young people complain that there is nothing to do unless you drink hard or are an addict of bingo. Vandalism is pretty ferocious, too, since too many kids are bored and aimless...". The category D status had prevented the village from getting a community centre; the Co-op, chapel and one of the pubs had all closed.

By the late sixties Durham's planners were starting to get impatient. The clearance process was seen as taking an inordinately long time; indeed only two or three small category D settlements had been completely cleared.

General view of the Morrison Busty Pit.
REPRODUCED BY KIND PERMISSION OF THE NORTH OF ENGLAND OPEN AIR MUSEUM, BEAMISH

However, although when the 1951 plan was first proposed only two of 31 district councils in Durham had raised objections to the policy, the inhabitants of the condemned villages as time went on were increasingly ready to make their views known. By 1967 a grassroots organisation CROVAC (County Redevelopment of Villages Action Committee) had been formed to coordinate the activities of over thirty village committees, to stop the category D policy and to press for the reinvigoration of villages. Its members argued that the county policy, assuming that villages could survive only as homes for miners within walking or cycling distance of their mines, was out-of-date; increasing availability of cars would allow the villages to function while inhabitants could travel to work elsewhere. They also felt discriminated against on other grounds: Barr recounted that "the planners' opponents say they have condemned (or reprieved) each village on essentially tidy-minded, *visual* grounds. If a settlement is compact, round, has perhaps an ancient Saxon church in the middle, it is okay. If it is just a few parallel rows of straggling terraces, a long thin place, it's not okay. They say that the planners are offended by the sheer number and scatter of Durham settlements, that they think there is something essentially meritorious about fewer, larger places…" Barr himself, although sympathetic to the plight of the inhabitants of 'D' villages, nevertheless saw the clearances as inevitable and basically to be welcomed. He concluded: "A way of life *is* being destroyed. But for all its virtues, it is not a way of life that can survive. On balance the planners are right."

The people of Quaking Houses, and many other places like it, had different ideas.

A vast question mark

A further threat to Quaking Houses' continued existence emerged in the 1960s with the development of the huge Chapman's Well opencast mine, one of the largest in County Durham, on its doorstep. This "comprised a deep excavation in the pattern of an inverted question mark drawn around the village", wrote Paul Younger, a water resources engineer at the University of Newcastle upon Tyne who later became a leading figure in the wetland scheme. "Naturally, the people of Quakies resented the noise, dust and traffic. (Although coal wagons did not come through the single main road in the village, bus-loads of gawking German tourists did, to stand in the bus-turning circle and marvel at the opencast pit)."

The Chapman's Well site was extended in the early 1990s despite opposition from local people (part of it remains in use as a landfill site, and a waste transfer station was also built on the area). Although the fight to stop the extension was unsuccessful, it did however lead to a significant development in the recent history of the village, with the formation in 1989 of the Quaking Houses Environmental Trust. The Trust was to prove an indispensable element in the development of the wetland project.

2 *The Sunderland scarf*

The Stanley Burn rises a few hundred metres to the west of Quaking Houses, emerging from the spoil tip for the Morrison Busty colliery which had grown since the sinking of the pit in the 1920s, on the boggy fields of Dogger Bank. It runs eastwards along a shallow valley to the north of the village, disappearing underground for a short stretch after passing beneath South Moor Road to emerge again near South Moor Catholic School. East of Craghead, the stream becomes known as the Twizell Burn; at Twizell Wood it is joined by another stream and it is this confluence that gives the burn its new name (from the Old English *twisla*, 'fork of a river'—the same element as in the Northumberland place name Haltwhistle). The Twizell Burn continues to flow roughly due east past Grange Villa and Pelton Fell and, now joined by the Cong Burn and known here as Chester Burn, through Chester-le-Street, its course—once again culverted and hidden from view—indicated by the street names North Burns and South Burns by the town's marketplace. At Chester-le-Street it enters the River Wear, which flows on past Washington to Sunderland and the North Sea.

The Stanley Burn's passage to the sea would in the past have been through a landscape studded by spoil heaps, the black Alps that dominated the scenery of so much of County Durham. The following description by J.B. Priestley of a spoil heap in the east of the county would not have been untypical:

> "This volcano was the notorious Shotton 'tip', literally a man-made smoking hill. From its peak ran a colossal aerial flight to the pithead far below… The 'tip' itself towered to the sky and its vast dark bulk, steaming and smoking at various levels, blotted out all the landscape at the back of the village… One seemed to be looking at a Gibraltar made of coal dust and slag. But it was not merely a matter of sight.

That monster was not smoking there for nothing. The atmosphere was thickened with ashes and sulphuric fumes; like that of Pompeii, as we are told, on the eve of its destruction." [J. B. Priestley, *English Journey* (1934)]

The landscapes of coal-mining—the vast smoking pit heaps—have, through remodelling or planting, been hidden from view or disguised as effectively as most other reminders of the industry. Priestley's apocalyptic description of the pit heap at Shotton, in East Durham, no longer bears any resemblance to conditions in and around Quaking Houses; but the present appearance of the Morrison Busty spoil heap, with its trees and grassy slopes facing towards Quaking Houses, belies its role as the villain of this story.

The Morrison Busty tip now occupies an area of around 35 hectares. It is composed of a typical mix of materials: shale, ash, coal and coal dust, with some cobbles, sandstone boulders, timber and red burnt shale, the spoil varying in thickness from 4.25 metres to around 11 metres.

In 1986–87 the A693 Annfield Plain by-pass was built, cutting across the area of the spoil heap, and the drainage system of the tip was incorporated into the design of the road drains. Water from the tip and surface water from the road was channelled to three outfalls, the most significant of which discharges into the Stanley Burn just a few metres downstream of its source.

Around the time the by-pass was opened, local people in Quaking Houses began to become increasingly aware of problems of pollution in the Stanley Burn. High levels of aluminium hydroxide in the water caused a build up of white foam—of the consistency of shaving foam—by the main outfall. High acid levels in the water caused iron to be deposited on rocks in the stream in the form of red and orange ochre (iron oxyhydroxide), while in deeper, stiller parts of the stream, the accumulation of aluminium gave a cloudy, milky appearance to the water; "a Sunderland scarf effect of alternating red and white segments", according to Paul Younger. A further problem came in the form of high levels of salt in the water, caused not by the spoil heap but by road de-icing salt in run-off water from the by-pass and from a nearby local authority compound where road salt was stored.

Local people believe that construction of the by-pass sparked off the pollution of the burn by disturbing the drainage systems within the spoil heap, although a 1995 report for Durham County Council by Joanna Markey-Amey suggested that the stream was probably polluted for some time before the by-pass was built, although its construction may have intensified the pollution process. What is clear is that the problem is caused by water bearing pollutants from the tip, rather than minewater *per se* bringing up pollutants from underground workings (this is also a serious problem across the north-east coalfields and one that has in recent years received rather more media attention than problems caused by spoil heaps). Open cast mining in the area has shown that the shallow High Main coal seam near the stream was dry, reinforcing the belief that the pollution of the burn comes from the spoil heap rather than from old workings.

Samples taken from the spoil heap showed characteristics similar to the untreated water of the Stanley Burn, containing iron and aluminium. The greatest concentrations of these metals, and the highest levels of acidity of water within the spoil heap were found close to the surface at levels of 0.3 to 1 metre in depth. It is thought that a substantial proportion of rainwater falling onto the spoil heap is 'trapped' in the surface layers by an impermeable or semi-permeable hardpan layer below the surface; this hardpan itself being formed as a result of the oxidisation of pyrite in the spoil, encouraging the formation of secondary minerals, notably geothite, which then fill the tiny voids between particles in the spoil, cementing it together and reducing its ability to allow water to percolate downwards. Some of the rainwater falling onto the spoil heap collects on top of this hardpan before flowing into the drainage system of the tip and the by-pass, and into the Stanley Burn; the bulk of the spoil drainage is via a thin saturated zone at the base of the heap, a process modelled by Newcastle University MSc student Shakwi Srour. Oxidisation of the pyrite in the spoil releases iron and sulphuric acid into the water; the acidity of the water allows it to leach aluminium and other metallic polutants from minerals in the spoil. This uninviting cocktail then passes into the drainage system and into the Stanley Burn.

Tests showed that the problem of pollution was greater in the summer than in the winter, in part because the oxidisation of pyrite in the spoil heap occurs more readily in warm weather, which encourages bacterial activity; and also because higher winter rainfall causes greater dilution of the contaminated material passing into the drainage system.

The result of all this was effectively to kill the burn, greatly impoverishing the local ecosytem for a kilometre or more downstream. "We found the Stanley Burn to be severely impoverished in terms of fauna", says Paul Younger, who first became aware of the problem in summer 1993 with the results of sampling by Kim Bradley. "There were only pollution-tolerant dipterans and oligochaete worms, but no fish of any kind". The deposition of ochre and the presence of aluminium in the water prevented photosynthesis from taking place in the stream. This reduced potential food sources for other forms of stream life. Insect larvae were largely absent; and, in a knock-on effect, impoverishment of stream life caused a noticeable reduction in the local bird population and in numbers of other creatures such as water voles. Additionally, the high levels of aluminium, iron, zinc and other elements in the water flow were highly toxic, and this had an effect on the ecology of the stream even downstream of areas where ochre deposition was a problem.

These very visible problems were duly noted by members of the local community. Individuals and members of the Quaking Houses Environmental Trust attempted to physically clean up the burn by removing debris and the more noticeable ochre deposits, but despite their best efforts the problem kept recurring. Informal observations of the state of the stream led to closer monitoring, attempts to involve other agencies, and eventually anger at the absence of any action to tackle the problem. According to Terry Jeffrey of the Quaking Houses Environmental Trust (QHET), "the way that we tended to monitor it was through people like Chas Brookes, who frequently goes out on walks around the area… Also, as we were cleaning the burn we began to realise that it was getting worse and worse. We managed to raise the problem of the pollution, starting off by making complaints to the local authority, the National Rivers Authority [forerunner of the Environment Agency] and Northumbrian Water about the state of it."

However, an increasingly vigorous lobbying campaign, approaching local authorities, MPs and MEPs, the NRA and other agencies and the local media in an attempt to get proper attention to the problem, only managed to elicit sympathetic inaction until March 1994, when Diane Richardson attended on behalf of QHET a meeting hosted by Durham County Council, to discuss the problem of minewater pollution in the Wear Valley. One of the speakers at that meeting was Paul Younger, who presented a paper on the effects of minewater pollution; one of the local examples he cited was the Stanley Burn at Quaking Houses, with analysis of samples from the burn. Diane Richardson approached him after the meeting explaining the QHET campaign on pollution of the Stanley Burn and its failure to obtain serious interest, and asking for a copy of Younger's findings. Two sides of the triangular relationship that led to the Seen & Unseen project had come together.

Quaking Houses

A pictorial record

Plate 1 Quaking Houses village from the east.

Plate 2 'Shaving foam' deposits of aluminium hydroxide at the outlet of the Stanley Burn from the Morrison Busty pit heap.
PHOTO: PAUL YOUNGER

Plate 3 The Sunderland scarf: red ochre deposits and water made cloudy by
aluminium give a red and white striped appearance to the stream.
PHOTO: PAUL YOUNGER

Plate 4 Gavinswelly: the pilot wetland at Quaking Houses.
Photo: Paul Younger

Plate 5.1 Diggers excavate the wetland site as Terry Jeffrey watches.
Plate 5.2 Getting stuck in! Volunteers help excavate the wetland site.
PHOTOS: PAUL NUGENT

Plate 6 The wetland under construction: the shape of the ponds and the PFA lining can clearly be seen.
PHOTO: PAUL NUGENT

Plate 7 Overview of the finished wetland.
PHOTO: PAUL NUGENT

Plate 8.1 A visit to the glassworks at Sunderland by local children.
Plate 8.2 A community planting session with young volunteers from Oxhill.
PHOTOS: PAUL NUGENT

Plate 9 The timber jetty across the wetland.
PHOTO: PAUL NUGENT

Plate 10.1 Peter Randall-Page, Shell Form, for Common Ground, Dorset, 1985-86.
Plate 10.2 Part of the slag dump at Nine Mile Run, Pittsburgh. A few trees begin to take root.
See Chapter 9.
PHOTOS: MALCOLM MILES

Plate 11 Lee Dalby, willow forms in the wetland.
Photo: Malcolm Miles

Plate 12.1 Willow forms in the wetland, by Lee Dalby.
Plate 12.2 Lee Dalby at a willow workshop.
PHOTOS: LYNNE OTTER

Plate 13 Willow forms in the wetland.
PHOTO: LYNNE OTTER

Plate 14 Installation at the Hancock Museum, Newcastle-upon-Tyne, by Helen
Smith.
PHOTO: PAUL NUGENT

Plate 15.1 Receiving the 1995 Environment Award for the pilot wetland.
From right: Terry Jeffrey, Paul Younger and (fifth from right) Diane Richardson.
Plate 15.2 Receiving the Henry Ford Award, 1998, at the Mansion House, London.
From left: Alan McCrea, Chas Brookes, Paul Younger, the Lord Mayor of London,
Helen Smith, Terry Jeffrey, Adam Jarvis, and the managing director of Ford UK.
PHOTOS: UNIVERSITY OF NEWCASTLE-UPON-TYNE

Plate 16 The Quaking Houses wetland in February 1999.
PHOTO: MALCOLM MILES

3 *Genesis of the project*

A number of very different influences, ranging from the personal and small-scale, to the influence of national political questions, led to the involvement of water engineers from the University of Newcastle in the Quaking Houses project.

A critical external factor had been the notorious November 1992 decision of the then President of the Board of Trade, Michael Heseltine, to close down a large number of Britain's remaining deep coal mines by 1995; this raised fears about future provision for the continued pumping of water out of disused mine workings, and led to minewater pollution and its effects being pushed up the political agenda and receiving serious attention from interested or relevant authorities and agencies, and from the regional media.

Dewatering systems in the form of pumping mechanisms had kept minewater at safe levels for the miners and prevented minewater pollution from seeping into surrounding streams and rivers. Inland minewater is caused by a build-up of rain water that has infiltrated the mines. In coastal coalfields, it is due to the ingress of both sea water and rain water. When pumping of this water ceases the water levels build up and seep out or overflow uncontrolled into adjacent rivers and streams. This minewater becomes polluted due to chemicals in the mine (mainly oxidised iron pyrites—iron disulphide) dissolving in the rain and/or sea water, followed by the reaction of these chemicals with oxygen as the water emerges from the workings. This is a problem in the southern and westernmost areas of the Durham coalfield, in western sections of the Yorkshire coalfield, and in Scotland and Wales. Where dewatering pumps have been switched off (in some cases 18 to 28 years or more ago) there is evidence that rivers in surrounding areas became polluted within a time span of approximately ten years.

Younger was aware that scientists are often guilty of scaremongering with what their research findings imply for the future. He admitted to wanting to stir people up about what would happen if the pumps were turned off in the coalfields, but also to do something that would raise hope that a positive solution to the problem may be possible. He acknowledged that there was a real problem and maintained: "I wasn't scaremongering; it's still true and they are going to see that it's true. It's already a fact in most of Scotland and Wales, so that activated me in a big way. I've got complex motivations."

Younger had been asked by the National Union of Mineworkers to look at the environmental implications of ceasing pumping of minewater on water resources in the River Wear catchment area. This prompted him to commence a study of minewater discharges in the west of County Durham where the mines had been closed for some years; information was requested from the National Rivers Authority that would provide him with sound comparative data to show what might happen if they switched off the pumps in the central part of the coalfields in County Durham. Because this information was not available, in April 1993 a research student, Kim Bradley, was asked to survey all the streams in the west of Durham close to pits which had closed in the 1960s, looking for the tell-tale signs of minewater pollution (high sulphate and low pH). A number of such discharges were identified, including the discharge into the Stanley Burn at Quaking Houses (this was only later recognised to be the result of spoil heap drainage rather than minewater effluent). This research and subsequent studies by Paul Younger and his colleagues in Newcastle University's Department of Civil Engineering, backed by correlating independent research by the NRA and lobbying by Easington District Council, Durham County Council, the National Union of Mineworkers and other organisations, combined to create enough pressure to ensure that the DTI would continue to maintain pumping at the nine regional dewatering stations in County Durham for the foreseeable future and that this responsibility would be passed to the Coal Authority when the Coal Board was privatised. It also made him an obvious choice to attend Durham County Council's March 1994 meeting on minewater pollution at which he met Diane Richardson.

But Younger's interest was more than purely academic. A native of County Durham, he was brought up in a mining village similar to Quaking Houses, and had an instinctive understanding of the ways of life of such communities: "For me the project, in allowing me to work with a mining community in County Durham, is 'the real thing'. I'm a scientist, but I'm from that back-

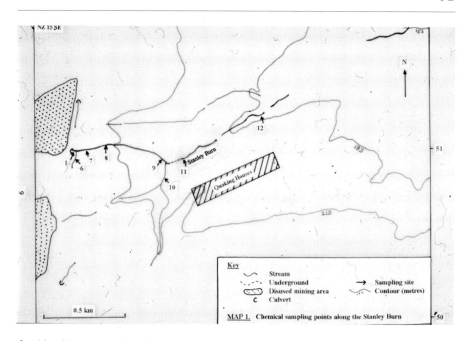

Quaking Houses and the Stanley Burn.
ILLUSTRATION: UNIVERSITY OF NEWCASTLE-UPON-TYNE

ground, and to put my scientific and engineering skills towards work of scientific merit for what I consider to be my people, my real live community—I'm not from that village, but I'm from the same sort of place—is a real thrill". He had also only recently—in 1992—returned to the United Kingdom after working in South America for some years for the United Nations Association, in particular working on problems of clean water supply for rural communities in the high Altiplano region of Bolivia. There, government intervention had been minimal and it was left largely up to individual village communities to plan and develop their own supply sources. These examples of self-help were to strike Younger later on as being applicable to the problems faced by Quaking Houses. It was not that the UK, and the north-east, lacked local authorities, agencies and bodies all charged with responsibilities and duties including the protection of the environment and the remediation of pollution; but, as he pointed out, "a string of letters… had received courteous responses acknowledging the problems, but (as is often the case) no funding had so far been available… to tackle the pollution". After the meeting, and his initial contacts with QHET, he reflected: "they went off and carried on with their letter writing and complaining, and I helped them out where I could. Then it just came into my head… all the work that I had been doing in Bolivia was community organisation based, augmented by people from outside who had other training and who could help the community to do things for themselves. In Bolivia, if you keep waiting for government money to do anything then you will wait for ever".

Younger was already involved as an external technical auditor with the NRA's programme to clean up the environmental damage at the former Wheal Jane tin mine in Cornwall, where the collapse of a seal in underground workings in January 1992 had led to an estimated 50 megalitres of acidic and metal-contaminated water spilling into the Carnon River and Fal estuary. There, although the scale of the damage was very different to that of the Stanley Burn (the ecological rescue plan at Wheal Jane has received over £16 million of government funding), some of the solutions adopted were ultimately to be applicable at Quaking Houses. Among these was the construction of a passive treatment system, the first to be planned in the UK (it was inaugurated in 1994), and based on parameters developed in the United States. This used the bioremediation processes of natural resources, with a combination of physical, chemical and biological mechanisms similar to those found in many natural wetlands. Such processes brought about the removal of metals and the neutralisation of acidity in the effluent water. Younger had also become familiar with the latest developments in this area in the USA, where 'passive' treatment of minewater discharges by passing the polluted water through artificially-constructed wetlands was also increasingly being trialled and adopted. "At the time there was nothing else like it in Europe", he remembers. "When I was first approached I didn't then know anything about passive treatment of minewater and had to become an instant expert on the subject".

In the Appalachian region of the eastern United States, many thousands of streams and rivers are affected by minewater drainage from abandoned pits. Until the mid 1980s, this pollution was—if it was treated at all—subject to chemical procedures to clean up the water. Typically, chemicals such as calcium hydroxide, sodium hydroxide, sodium carbonate or ammonia were added to the water to neutralise its acidity and to cause the metallic toxins present to solidify out of the streamwater, which was passed through sedimentation basins where solids would settle. This process also often required the introduction of secondary chemicals, to aid oxidisation and the coagulation of pollutants, and mechanical devices such as aerators and mixers.

Chemical treatment is reliable but is also expensive, both in infrastructure and annual running costs, and not surprisingly for very many smaller discharges the pollution was tolerated as any treatment options were considered unrealistically expensive.

However, in the mid 1980s increasing attention began to be given to the passive treatment of acidic minewater pollution after observation of the effect on the water quality of minewater discharges flowing through naturally occurring wetlands. Interest in a number of pilot projects for artificially created wetlands, highlighted at a 1988 conference on mining and reclamation, led to an increase in wetland construction in the Appalachians. Typically, these efforts have been spearheaded by stream restoration groups involving local universities, conservation organisations, landowners and sporting interests, often in partnership with local industries, charities and state reclamation bodies and federal sources of funding which have provided around $10 million over five years for projects in the region. At first, results were extremely variable and the initial optimism about this 'new' technology gave way to an over-reactive pessimism. But by the mid-1990s the collation of relatively long-term results for a number of projects and the development of conceptual models to assess the viability and inform the design of wetland schemes has allowed a more pragmatic view to be taken of the use of passive treatment technologies.

These developments were at the forefront of Paul Younger's thoughts as he pondered the problems of Quaking Houses. "Could a wetland be the answer at Morrison Busty?" recalled Younger. "Having discovered two possible sources of funding for community-based research, I drove up to Quaking Houses in search of the Environmental Trust. True to form, I had forgotten Diane's name, and had never noted her address. Hence I wasn't sure how much luck I would have in finding the Trust. I needn't have worried. The first person I stopped to ask for advice knew exactly whom I needed to contact, and exactly which letter box I needed to drop a note through. Within twelve hours, Diane called to say the Trust were keen to support a bid for funding, and I began writing in earnest".

Younger had written to Diane Richardson saying that he was definitely interested in investigating the issue of minewater pollution in the Stanley Burn. His initial aim had been to create a wetland at Quaking Houses which the University could use as a teaching resource, and he was also hopeful that money might be made available from a university fund endowed by the novelist Dame Catherine Cookson if, as Younger said to Diane Richardson, "…would [Trust members] be as keen to pick up spades as you are to pick up pens?… If I could get money for materials would you be prepared to do physical work?" Quaking Houses residents were more than prepared to back up their wishes with hard work, and a formal bid was submitted to the Catherine Cookson fund, only to be turned down; as was a similar bid to Earthwatch. These failures to secure funding were for the most frustrating of reasons, one of the rejection letters stating that: "We liked your proposal very much. It is good science and good community work. However, we feel that there ought to be public funding available for this sort of thing". "I couldn't have agreed more, there ought to have been public funding available, but up till then, there simply hadn't been", commented Younger.

Meanwhile, the Trust's energetic lobbying was beginning to pay dividends. At a meeting with the local Euro MP Stephen Hughes, members had raised the issue of the pollution of the Stanley Burn, and Hughes had written to the Chief Executive of the NRA protesting at the length of time the situation at Quaking Houses had been allowed to persist without action. This led directly to the NRA making funds available for a feasibility study of the options for treating the pollution in the Stanley Burn, and one condition in the invitation to tender for the project which the NRA subsequently issued was that the successful candidate should liaise closely with the Trust to ensure that any measures taken were supported by the local community.

An initial £13,000 was made available by the NRA, and Newcastle University's Civil Engineering Department was awarded the tender for the work, which involved two studies: of the Stanley Burn at Quaking Houses and of a fluorspar mine in the North Pennines which was giving rise to zinc pollution of a local waterway. Younger sought advice and contributions form a number of scientists, each representing different areas of the discipline. In addition, he also employed a number of research assistants who contributed to the project in different ways. Principal among these, and involved most closely both in the feasibility study to build a pilot wetland and, eventually, in Seen & Unseen, was Adam Jarvis. In 1994 Jarvis was subcontracted for a six-month period to carry out the feasibility study that Younger, in collaboration with QHET, had won the bid for. Initially the objectives of Jarvis's post, in consultation with Younger, were to determine the extent of the discharge, its cause or causes, and to investigate if and how it could be treated, and whether a pilot wetland was feasible.

NRA agreed to fund an extension to the study—originally funded only for three months—lasting until March 1995; this permitted Younger and his colleagues, and the Trust, to design and construct a small passive treatment wetland near the outflow at the base of the Morrison Busty tip. On 20 February 1995, volunteers from Quaking Houses, scientists from the University and NRA officers began to dig at the site, using just hand tools and elbow grease. Construction of the Gavinswelly wetland, which was to be a precursor of the eventual Quaking Houses wetland, had begun in earnest.

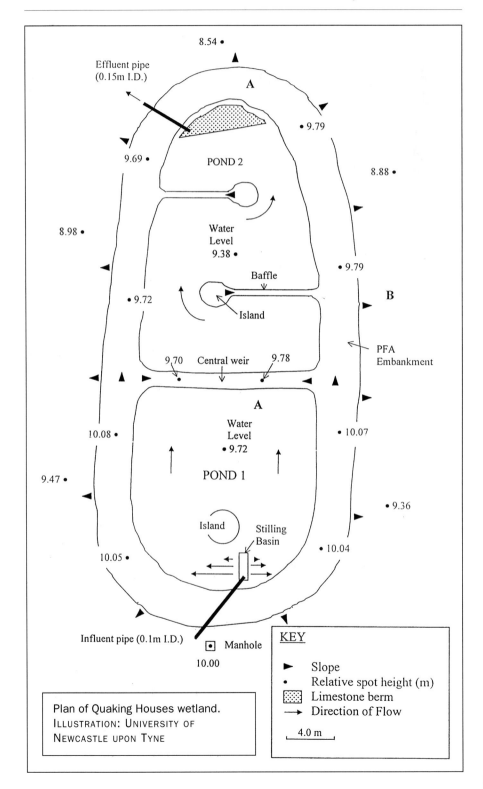

Plan of Quaking Houses wetland.
ILLUSTRATION: UNIVERSITY OF
NEWCASTLE UPON TYNE

4 *Gavinswelly and beyond*

Why the Gavinswelly wetland? All place names have a meaning even if— like Quaking Houses itself—that meaning is somewhat obscure; and Gavinswelly is no exception. Picture the scene: 20 February 1995, a brisk winter's day, and a group of stalwart volunteers—from the Environment Trust, University, NRA and local children—digging with a will at the site of the pilot wetland. Digging in rather boggy conditions, too, hence the notice erected on site: 'Gan Canny! Claggy Clarts'. Just then, as Paul Younger recounts,

> "Young Gavin discovered that he was himself the first victim of the claggy clarts. He couldn't remove his leg from the deeper, wetter portion of the excavation. We tried all sorts. We pulled and pushed and dug and scraped, but his foot wouldn't budge. At least not with his welly on it. After some serious deliberation, we decided that the welly would have to be sacrificed in the name of science, to be buried beneath the wetland for posterity. Gavin accepted this decision with good humour, notwithstanding the prospect of a half-kilometre walk home in his stockinged feet. Nevertheless, to lessen the blow, we decided to commemorate his sacrifice by naming the pilot wetland after his ill-fated welly. And so the Gavinswelly wetland was christened."

Even more importantly, what *was* the Gavinswelly wetland?

Based on their researches, Younger and his team had decided on a compost-based wetland as the best solution to the problems posed by the Stanley Burn. The pilot wetland was an anaerobic (not requiring exposure to free oxygen or air) compost wetland, composed of four parts, or cells, in series. The first two cells contained saturated horse manure and soil; the third cell contained open water and limestone; the fourth, an aerobic overland flow system established in *in situ* soil. In the first two cells, sulphate-reducing bacteria living in the compost reacted with the sulphate-rich minewater flowing through, producing sulphite radicals which combined with the iron to form insoluble precipitates which settled on the bottom of the cell. Iron sulphide is a benign material as long as it is kept isolated from oxygen; otherwise, it would tend to increase the acidity of the water. As part of the process, the water became less acidic and more alkaline, causing the aluminium present in the water flow to separate out as aluminium hydroxide and also to settle in the compost substrate at the bottom of the cells; the aluminium hydroxide over time settles and becomes gibbsite, an innocuous compound found in most soils. As the aluminium separates out of the water, its acidity rises again, so the water is then passed over limestone chippings in the third cell to raise its alkalinity to a state approaching normal levels for the area.

This is a very different process from the relatively better-known form of passive water treatment, aerobic wetlands, commonly described as 'reed beds', used to treat organically polluted water, for example at small sewage works. While reed beds achieve the breakdown of organic matter through surface flow of the polluted water over shallow gravel beds planted with reeds, the anaerobic processes used at Quaking Houses were to remove dissolved metals from the water, with most of the water flow occurring below the surface.

Adam Jarvis explained: "At the time, passive treatment was coming to the fore as a method of treatment because it was cheaper than the alternative active treatment, and, because of the legislative framework, cheap solutions were by far the most attractive. The pilot scale wetland was very efficient. It removed lots of the metals and acidity—which are the critical contaminants." The pilot wetland treated about 10 percent of the minewater. Iron and aluminium concentration dropped by about 80 percent in the treated water. In addition to removal of these chemicals a diverse range of wild fauna were seen using the newly cleaned water at the wetland, including several species of water beetle and water-boatmen, newts during their breeding season, and wild deer who used it as a source of drinking water.

The pilot had proved astonishingly successful. Younger and Jarvis had expected that it would take a settling-in period of some months before the full benefits were apparent, but in fact a substantial improvement in the iron and aluminium content of the water flowing out of the wetland was apparent

'Gan Canny. Claggy Clarts.'
PHOTO: PAUL YOUNGER

almost immediately, as was a decrease in the acidity of the effluent. One problem that had been anticipated was that the compost-rich wetland would generate such a high biochemical oxygen demand (BOD) that it would cause a severe problem of oxygen depletion in the effluent, thereby hampering the ability of the cleansed water to support life. This problem failed to materialise in any significant form. When the pilot scheme was first commissioned, the effluent did indeed have a very high BOD of around 800mg/litre, but within two weeks this had declined dramatically to just 2mg/litre, as the more mobile organic substances were washed out of the compost layer. Performance of the pilot wetland was monitored weekly for the next year and a half, until the summer of 1996.

The pilot wetland was the first anaerobic wetland area treating minewater pollution to be brought into use in Europe—the Wheal Jane scheme, although planned earlier, took longer to be completed. Quaking Houses differed in another crucial respect: although the chemical processes were anaerobic, taking place below the surface of the water, the wetland was open to the air. At Wheal Jane the anaerobic cells took the form of sealed tanks, covered with liner and clay. Another scheme developed after Quaking

Houses at a site near the River Pelenna in South Wales, while open to the
air and with planted reeds as well as a compost substrate to provide both
aerobic and anaerobic treatment, took the form of 'wetland' contained in
rectangular concrete tanks. But the Quaking Houses scheme was designed to
be part of the landscape, a concept which became even more important once
the decision was taken to press on with the development of a full-scale
wetland. "We wanted a more natural system, with growth and decay and the
continual adding of organic plant matter to the wetland", commented Paul
Younger, who described his vision of the wetland (albeit in the context of the
way in which the work of the artist would affect the site): "I want people to
look at it and be fooled into thinking that it has always been there, not immediately but perhaps in eighteen months time… I don't want them to walk up
to the site and say that here is a completely manipulated thing".

The success of the pilot wetland led the University and the Quaking
Houses Environmental Trust to seek ways of developing a full-scale wetland
able to treat all of the water flowing into the Stanley Burn from the Morrison
Busty spoil heap—a flow which ranged from 17 litres per minute to over 100
litres per minute. This new phase of an already ground-breaking project was
to bring in the third side of the triangle—the employment of artists and the
close involvement in the running and direction of the project by the Artists'
Agency, as described in the next chapter.

In technical terms, the full-scale wetland as finally designed was similar in
concept to the pilot, although this had not actually been intended by Younger
and Jarvis from the start.

At first, they planned to install a vertical flow Successive Alkalinity
Producing System at the site, which is believed to require 40% less land area
than the more normal horizontal flow system to handle a given amount of
water. However, a detailed site survey revealed that the site would only allow
a two-metre head of water, far less than would be required for the SAPS
design. It was also found that the site had formerly been an unrecorded finings
pond of the Morrison Busty pit, and the land was highly contaminated with
iron pyrite; on exposure to air this would oxidise and be rapidly washed into
the stream water. So on-site excavation had to be kept to a minimum to avoid
polluting the Stanley Burn still further.

This major site survey had not been possible before the appointment of the
first artist to the project, Jamie McCullough, because the Lankelly Trust who
were providing part of the funding would not allow any work to be under-

taken on the site until all the funding was in place and the artist had been appointed. Small-scale survey work had been undertaken but this major survey required some idea of how the site was going to be used before it could be decided how much material to remove and also because specialised digging equipment would have to be hired and costs met to form the major budget for the wetland construction. The survey revealed that the site relief was higher than they had originally thought—only two metres difference in height between the source of the minewater discharge and the proposed exit point from the wetland. Together with the problem of the contaminated soil these results had implications for resources, requiring increased financial outlay and imposing major design constraints. Three designs were considered—two continuing to incorporate elements of the SAPS design—but in the end it was decided to go for the horizontal flow option.

This unexpected hiccough—which also precipitated the departure from the project of Jamie McCullough, whose relations with the scientists and the Artists' Agency had become increasingly difficult—had caused construction of the wetland to be delayed well into 1997. Construction eventually began on 14 July 1997 and it was largely complete by 11 September 1997. Covering approximately 300 sq m, it initially consisted of two linked cells—rather, ponds—to which a third pond to allow extra cleaning of the water has since been added. The ponds are around 30-50cm deep, enclosed by embankments made of 800 tonnes of pulverised fuel ash (PFA) supplied by National Power, a waste product from the electricity power generation industry which is increasingly recycled for construction uses. The first, upper pond receives the minewater discharge via a 4" pipe and stilling basin, and contains one island to aid the distribution of flow. It contains 60 tonnes of cattle dung, horse manure from a local stables, and compost derived from municipal waste. The stream water then flows over a central embankment into the second pond, which contains two baffles and islands to encourage a circuitous flow pattern. At the end of this pond is a berm or embankment made up of 25 tonnes of limestone (a further 5 tonnes is distributed across the two ponds). The water now flows from this pond through the final, smaller pond and returns to the Stanley Burn.

The active ingredients—the compost and limestone—are expected to last for around 20-30 years before needing to be replenished (although it is interesting to note that had limestone been used in isolation to correct the acidity of the water, it would rapidly have become coated with aluminium and iron hydroxides and would have lost much of its effectiveness in restoring the pH balance of the water). The construction of the wetland cost around £12,000, comparable to the minimum capital cost of an active intervention treatment facility, but also equivalent to the annual running costs of such a facility, whereas the wetland has negligible running costs.

The wetland was commissioned on 10 November 1997 and the first water samples were taken on 14 November. Initial results were very encouraging, showing reductions in iron, aluminium and even manganese concentrations in the treated water, and a decrease in acidity. The results of the first samples tested showed the pH of the water rising from 5.27 on entering the wetland to 6.75 on leaving it (pH7 is neutral; below 7 denotes acidity, above 7 alkalinity). They also showed that iron concentrations had fallen by three-quarters, from 20.5 mg/litre to 5 mg/litre, aluminium content by nine-tenths, from 10.9 mg/litre to 1.1 mg/litre, and manganese levels halved, from 5.5 mg/litre to 2.8 mg/litre. More detailed scientific and statistical findings are contained in the Appendix.

5 *Visions of utopia*

Innovative projects require people of vision. Innovative, interdisciplinary, collaborative projects require not only vision but respect for each others' discipline. Add in a dedicated, willing and receptive audience who actively wish to play a participatory role, plus a series of fortuitous events, and you have the ingredients for a new and exciting venture, the Seen & Unseen.

Lucy Milton, Co-Director of the Artists' Agency, had the belief in an ecological collaborative project based around water. Dr. Paul Younger from the University of Newcastle had the expertise plus a willingness to see the artistic input into what could have been a simple scientific solution. The Quaking Houses Environmental Trust (QHET) sought eradication of the problem of minewater pollution of their river, the Stanley Burn, and were prepared to work with both the artistic and scientific community.

This culminated in the building of a wetland, using a natural regeneration technique involving waste, with a fully accessible public route, including a boardwalk across the wetland, which linked with the existing footpath networks in the area. Alongside the practical building works, a programme of educational work in local schools and colleges was developed. Working with the local youth club and the QHET, a series of radio broadcasts were made and a world wide web site was created.

Throughout this project, the key players kept their faith despite some major difficulties. Valuing the contribution made by each of the disciplines placed a strain on the relationships at times and delays in completion and lack of funds were major headaches. It is to everyone's credit that this project overcame all these problems and the two national awards the wetland received are testament to the vision of those involved.

The early days

The QHET is one of a growing number of effective, articulate and extremely successful campaigning teams. They have been fortunate to be represented at different stages throughout their projects by Diane Richardson, Terry Jeffrey, Chas Brooks, Alan McCrea, Maureen Wilson, Tommy Cole and Douglas Glendinning, all articulate, forceful personalities determined to do the best they could for their small village.

A number of projects were undertaken that led to significant improvements in the area and these in turn led to a greater understanding by the whole village of the need for environmental protection. A community garden, partly designed by the village children with professional assistance, won a Shell Better Britain award and awakened the villagers to the involvement of creative outside agencies. Following other successful community projects, the QHET turned their attention to the effects of minewater pollution. Over a period of time, the QHET made various attempts to clean up the burn by removing debris as well as scooping up the red or orange-coloured deposits that had gathered on the banks and streambed. It seemed that no matter what they did, the pollution always returned.

Members of the trust began to worry about the effect on their children's health when playing in the water. They were eventually told that the water was not safe and they should keep the children away. Terry Jeffrey, a key member of the QHET, began to keep a watchful eye on the level of pollution and an active campaign began to find a solution. Members of the QHET had researched the problem and discussed the possibility of creating a wetland as a solution but it was not until they met Dr. Paul Younger that they had the scientific expertise to investigate seriously the creation of a wetland to combat the pollution. In any case, QHET felt it important to keep an open mind about of the causes of pollution, although Terry Jeffrey was convinced it was either the mine or the large landfill site that had grown up behind the village.

The Artists' Agency, founded in 1983 by Lucy Milton, seeks to bring artists in direct contact with communities, public services, social groups and industries by means of placements. Lucy Milton describes their aim as "offering artists an opportunity to extend the scope of their work in a new environment; to facilitate an exchange of ideas, knowledge and skills, leading to a greater appreciation of the arts; and to encourage individuals to develop their own creative skills."

Lucy Milton had believed for a long time that the artist had a pivotal role to play in helping to combat environmental degradation. She readily acknowledged that her approach and commitment to the future project had its roots in her own personal concerns about global warming and the effect that climate change may have on the future, in particular her daughter's future. She began by thinking about how an individual could improve the environment for future generations:

"I thought; okay it's vast, it's big, I would be stupid to think that I could have any input. Then I thought, but if you look at it on a micro level, you as an individual have knowledge and skill in one area and mine is in the arts. If you with your skill can do something in a local place then perhaps you can link up with other people who are doing small-scale things in local places."

These initial thoughts were turned into positive action. Through personal study of different environmental issues and dialogue with individuals and representatives from a variety of different organisations, Lucy Milton decided to concentrate her efforts within the remit of water pollution issues. She set up a research project that considered the possibilities for arts-based projects relative to the theme of water pollution within her own locality (the Northern Region of England). She believed that through networking and sharing of experiences the research project would reveal "potential to address ambitions to develop dynamic new ways of exploring the subject (water and the environment), and to explore the potential for effecting conceptual and political change through art."

She produced a paper that outlined her wish to involve artists in community environmental action under a project entitled the Seen & Unseen. Water pollution, she felt, was the perfect candidate for a collaborative, interdisciplinary project which could bring together both the artistic and scientific communities to combat environmental degradation.

It was at a Millennium Commission event to host the Visual Arts Year in the Northern Region that Lucy Milton circulated examples of her ideas for projects by Artists' Agency, including a project based around the theme of water. "Water is one of the universal mythologies throughout the world and as such is a powerful metaphor for life. Although it is essential for life on this planet it is generally taken for granted in western cultures where it is greatly wasted and abused. Elsewhere the effects of man's interventions are often disastrous, causing cyclones, floods and droughts. On a larger scale water can be used for highlighting the issue of whether the world is capable of sustained development or whether ignorant profligacy, greed and short term self interest are currently sowing the seeds of an environmental catastrophe."

Jozefa Rogocki, an artist from Darlington, attended the meeting and read this. She felt an immediate affinity with the project and was delighted when some time later Lucy Milton asked her to join the project as co-ordinator on a 1½–2 days a week basis and to undertake a feasibility study into the project.

A research grant of £2000 had been awarded by Northern Arts enabling Artists' Agency to develop the study, a key element of which involved hosting a brainstorming session in September 1994 that would bring together people from different disciplines to look at the issues.

Bringing together the key players

The 'brainstorming day' was held at the University of Newcastle on 19 September 1994 and hosted by the Civil Engineering Department. A number of organisations representing the different disciplines were invited to participate. Two different perspectives that proved to be central to the development of Seen & Unseen were presented by the multi-disciplinary arts-based ecological group PLATFORM, who hailed from London, and by Dr. Paul Younger and his scientific team from the University of Newcastle.

PLATFORM presented a project entitled 'Still Waters' which focuses on the loss of rivers in London. Through this they introduced a model of collaborative and interactive art practice that many people in the audience had never encountered before. One of the group from the scientific community later said, "For me, as a scientist and engineer, art is drawings, so it hadn't occurred to me that there's practical things…"

Participants in the day's event were clearly impressed with the focus on "thinking locally and acting globally" that PLATFORM exemplified.

Dr. Paul Younger's paper, entitled 'Topical Issues from a Regional Perspective', described his work on ground water pollution. He began his paper by stating that "the universality of what we talk about is neither here nor there" and that "regional is only important because we are here." He argued that topical doesn't necessarily mean popular and that by describing a site as topical we don't necessarily make it unique.

He stated, "There is nothing unique about County Durham, except that for those of us who live here it's unique, it's the place where we live."

During the presentation he highlighted the problems scientists often have communicating to others the seriousness of groundwater pollution. His presentation focused on a topical issue in the locality, minewater pollution in County Durham, and highlighted the problem of generating an awareness of the issue because it is not always visible or it is only visible in places that are inconvenient to get to. Recognising the relevance of the Seen & Unseen dimension, it is, he said, "the problem of making the unseen seen... We are at a loss at times as to how to get the images and ideas across most effectively... This is part of the political process."

He then added that what he found most exciting about Artists' Agency approaching him was that it created an opportunity to reveal things that were once unseen:

"The whole idea of the Artists' Agency Water Project is that it can help to illuminate or illustrate these (hidden) things. As scientists we use a lot of visual things but usually a lot of creativity doesn't go into it. We use standardised forms."

One example, which Paul Younger referred to, was the village of Quaking Houses in the west of County Durham. He described how he wanted to reveal "sources of hope and optimism" by focusing on the problem in this particular village. He stated,

"It's the sort of thing that keeps you awake at night once you start thinking about the details of it, the idea that a small community themselves, who are affected by these ideas, could do something about it and construct a wetland that hopefully provides some sort of diversity of habitat in what hitherto had been a completely dead stream would be a source of hope. All be it small, it's something."

He then moved onto the issue of funding and suggested that, in looking to PLATFORM as an example, there might be scope for joint funding alongside Artists' Agency. The funding would not only be of direct benefit to Quaking Houses but would be a source of hope and optimism for the region.

He explained,

"So there are possibilities here, of working with a community that is affected by minewater pollution. The importance of that would not just be for Quaking Houses but for hope and optimism. A lot of people like myself are very anxious indeed about the future of the coalfield itself and if these pumps are switched off then it is seriously going to affect the water supply for Sunderland."

Of the artistic role in the project he concluded:

"There is a role for the arts in trying to make people realise that biodiversity is of far greater merit in terms of the long-term viability of the ecosystem in which we live than we can ever hope to quantify in terms of economics. People with creative skills could perhaps find methods of articulating the complexity of the problems that face scientists working in this area by stripping them down to reveal areas of importance."

It was this event that marked the coming together of many of the key participants in the Seen & Unseen project.

The collaborative process begins

Following the brainstorming day, Paul Younger felt it appropriate to introduce the members of the QHET to Artists' Agency and the seeds of a collaborative wetland project were sown. Paul Younger was already working with the QHET on the development of a small-scale pilot wetland, managed by his colleague Adam Jarvis. It was immediately recognisable that here was a project which could fulfil several objectives. Lucy Milton's dream of an interdisciplinary ecological project; QHET's determination to combat the pollution in the Stanley Burn; Paul Younger's wish to use his scientific expertise in a practical and imaginative way to help a local community.

The next step was to find ways to move forward to make the process become an actuality. The key players decided to create a support group for Seen & Unseen which could oversee the management of the project.

The support group and the search for funding

A support group for Seen & Unseen was formed by Artists' Agency and QHET were invited to join alongside others including Martin Weston, the Arts Officer from Derwentside District Council, Julie Ross from Glasgow School of Art and members of Paul Younger's department at the University of Newcastle.

With the support group in place, minds needed to focus on the task of finding funds to start the project. Generating the funds to build the full-scale wetland proved to be a difficult and slow task. Both Lucy Milton and Paul Younger were determined to find a way of facilitating its construction. Paul Younger admitted that his motivation went beyond the demands or needs of his department. He was also motivated by a personal desire as he, himself, was brought up in a small mining community, not unlike the village at Quaking Houses.

Following the brainstorming day, a visit to Quaking Houses was arranged to discuss the potential for a collaborative project involving the arts, sciences and the community. Lucy Milton, Jozefa Rogocki, Paul Younger and Julie Ross met with Terry Jeffrey, Diane Richardson, Alan McCrea and Chas Brookes of the QHET.

Lucy Milton introduced the project from the perspective of Artists' Agency and talked about her personal concerns about the environment. Jozefa Rogocki showed some examples of American artists who had worked on environmental projects. Paul Younger revealed how it was not until he had heard the presentation from PLATFORM at the brainstorming day that he made the connection between art and science in a practical collaborative project. He had gone to the brainstorming day prepared to talk about minewater pollution and amongst his slides on the subject were some images of the pollution at Quaking Houses.

He explained:

"Then when I saw what PLATFORM were on about … it came to my mind when I was talking about the minewater that I ought to be talking about Quaking Houses. The idea of doing something on the arts side that's practical and for the community … so it seemed like the sort of thing that I had been trying to pursue through other channels. I mentioned it and just about everybody who spoke at the brainstorming day was in favour of it."

Following a tour of the village, and a walk along the Stanley Burn to the source of the pollution, it was agreed that Artists' Agency and QHET would work together with the scientists. A priority was to generate funding to appoint an artist to work with the local community on improvements to the environment linked to the construction of the wetland. It was agreed that Artists' Agency should lead on the fundraising, and keep everyone else informed of the progress. QHET contributed information about the village, and Paul Younger about the scientific aims.

Lucy Milton, as Co-Director of the Artists' Agency, agreed to take on the role of fundraiser, but said that she would value any support she could get from QHET. She pointed out that this would take a period of time to achieve and she could not guarantee any results. It was agreed that regular contact should be maintained amongst the support group and that it should meet at least once every three months. It is thanks to Lucy Milton's fundraising skills that an assortment of funders known to support the arts and sciences slowly came on board. Northumbrian Water's Kick-Start Scheme contributed £54,000, which in turn enabled a matching grant from the Pairing Scheme, and Northern Arts gave major support for the development stage of the project. Organisations such as The Arts Council, Shell Better Britain, The Rural Development Commission, The Gulbenkian Foundation and English Nature agreed to commit funds to Seen & Unseen. However, with each new funding stream came new project deadlines and budgeting complexities, the management of which becomes an art in itself. The group, which had a commitment to making connections between the local and the global, had much soul searching over funding offered by Shell and its Better Britain campaign, when they considered its perceived record on environmental performance, especially overseas. After a great deal of discussion, they helped develop a linked project, 'Funding for a Change', led by PLATFORM and AN, which sought to raise awareness around ethical issues for small organisations accepting corporate sponsorship.

The support group became an evolving collection of people who were keen on the project but looked at Seen & Unseen from many different angles. The diversity and interweaving of these organisations created a rich organisational texture. And the strong personalities at the centre of the project led to a determined and enthusiastic organisation, albeit sometimes with very forth-right views which made consensus-seeking difficult at times.

The funding made it possible to appoint Adam Jarvis, the scientific research assistant based at the University of Newcastle who under Paul Younger was involved in the feasibility study to build the pilot wetland, to work on the collaborative project to build the full-scale wetland. The scientific input was in place and the harder task of appointing an artist for the project began.

Whether it had been thoroughly appreciated at this stage, the differing views on what the artist should contribute are difficult to access. As we shall see, and it is worth spending some time on the appointment process, some very different ideas emerged as to the role of the artist in Seen & Unseen.

Appointing an artist

During the period between the inception of the project and the appointment of the first artist to be involved in the construction of the wetland, a variety of different perspectives were held by various members of the support group as to what they felt the artistic input would involve. Artists' Agency aimed to organise a project that would allow sufficient time for an artist to "produce work of calibre" and that the artist would have access to the scientists all year round. The main sponsors wanted see a set of designs by the end of the financial year (1996-97) and an attractive, functioning wetland at the end. The villagers wanted a clean stream and attractive surroundings. All agreed that access was an issue and that a boardwalk or bridge design allowing for disabled access ought to be included.

Martin Weston, the Arts Officer from Derwentside District Council, had assisted Artists' Agency with projects in the past and was therefore familiar with their ethos and practice. He indicated that the council would want the artist to make a positive contribution for future generations and to work with the young people in the village.

Although a consensus was reached about timescales and promotional events, personal aspirations about the anticipated artist and what they may achieve were varied. This was not surprising since the support group was entering new territory and each member was aware that the project relied on a collaborative approach embracing as far as possible consensus decision making. Even Artists' Agency had not had the experience of placing an artist in an ecological, interdisciplinary, collaborative project, and therefore translating the theory into practice was a new experience for all.

The following statement from Terry Jeffrey gives a glimpse into the thinking of some members of the QHET and residents in the village:

"It didn't seem like a crazy idea to work with artists, not at all. I think that the lads couldn't quite get on board with it, but I think that Diane (Richardson) didn't mind what happened. Basically, we weren't at all worried, we were concerned about how will we clean the burn? What will happen? That was basically the level that it was at. That was that."

On a more personal level Terry Jeffrey also responded, "I personally was really excited and I always refer back to C. P. Snow's book, *The Two Cultures*, where he talks about art and science and how they can both work together. So, I was on board with this concept, and I thought that this was brilliant."

Artists' Agency expected the artist to work with the community on both the construction and interpretation of the wetland. Following Martin Weston's view in particular, the artist should work with the young people in the village who represented investment in the future. For Seen & Unseen to work, collaboration between the artist, scientist, community and project evaluator was essential.

However, differing views on the role and function of the artist were emerging from members of the support group, which was not surprising since all participants were involved in an innovative project that required a new way of working based on interdisciplinary collaboration…

Paul Younger, the senior scientist on the project, felt it important that the artist was a community-centred person. "We are making a big investment. Whoever the artist is, it's got to click. We might get someone who is really good in interviews but really bad with kids or something."

Paul Younger was of the opinion that the artist would bring lateral thinking to the project by forcing the scientists, on the more technical side of the project, to reconsider ideas and become involved in the more permanent side of the work. He did express concerns that the project might grow too quickly and become pretentious. He wanted a wetland that functioned scientifically.

The support group noted "that the artists' work was likely to be organic rather than structural. Low maintenance should be mentioned as essential in the job brief, and there would need to be a separate maintenance provision for this work."

At a meeting on 2 March 1995, the support group decided that "we were not looking for an artist who would work purely as a designer or landscaper, or creator of public features; rather as an artist who would work in a much more experimental manner, and where the energy of the wetland could itself, for example, be a metaphor for regeneration."

The final brief for the artist was co-written by the team and stated that the artist would "explore innovative approaches to the creation of the wetland" and that the artist would be given one month's research time at the start of the project to explore experimental approaches such as interdisciplinary collaboration. It was envisaged that the artist would want to develop an ongoing relationship and undertake research within both the university and the community as soon as possible after his or her appointment. It was agreed by the support group that the artist would "want an ongoing relationship and an intense collaboration with the scientists." The post was open to artists working in any medium and individuals and groups of artists were welcome to apply.

The original plan was to interview and appoint an artist by late 1995. The scientists' engineering input would be resolved by the end of the year and the wetland construction would be under way by early spring 1996.

Artists' Agency provided a guiding framework for the selection of an appropriate artist. This framework was flexible and all decisions were made at the support group meetings. The post was advertised in the *Artists Newsletter* and *The Guardian* and fifty-two applications were received, each submitting visual work and a comprehensive curriculum vitae. Artists' Agency gathered the applications and it was agreed that the support group would meet at Quaking Houses to short-list four to six candidates for interview. Short-listing was to be based on the written materials that the applicants had submitted and each member was asked to compile lists of suitable and unsuitable candidates before coming to the meeting. At the short-listing the group was able to view the visual material produced by the candidates.

The short-listing took place on 6 June 1995. Lucy Milton and Jozefa Rogocki from Artists' Agency, Paul Younger and his research assistant Adam Jarvis, Terry Jeffrey and Diane Richardson from the QHET, Martin Weston from Derwentside District Council and Julie Ross, the project evaluator, were present.

The six candidates that were short-listed represented environmental, conceptual and site-specific work and came from a range of fine art disciplines including printmaking, painting, sculpture and public art. Members of the support group interviewed the artists. The final interview panel consisted of Lucy Milton, Jozefa Rogocki, Diane Richardson, Terry Jeffrey, Paul Younger and Peter Woodward, who represented Shell Better Britain, one of the project funders.

For Artists' Agency the selection of artists with a community group is part of their everyday work and they were relaxed about the interview process. It was a little more difficult for other members of the group to evaluate the artists' applications and a completely new experience for the representatives of QHET. The scientists approached the whole process in a very rational manner, looking for examples of visual work that would 'fit' into their wetland design. Artists' Agency focused on the potential and imagination in the visual work presented, attempting to pick up on artists who would be flexible and adaptable in their approach.

Everyone was given an opportunity to air their views and, after much debate, a consensus was reached.

The panel agreed that Jamie McCullough, an artist from Scotland who had been working on stream- and water-related projects, be invited to join the collaborative team. To be fair to Jamie McCullough, he made it clear from the outset that he wanted to do the initial research work away from the community and that elements of the project, in particular the community-related work, were not his main interest. At that time it was felt that, if necessary, other professionals could be brought in to make up any shortfall in Jamie McCullough's approach. At his original interview Jamie McCullough had said,

"It bothers the hell out of me setting up all the publicity, major publications, internet, etc., before the start. It tends to force you onto the safe, boring path that guarantees a result however trivial, instead of the risky, wonderful thing that might fail. Besides, you need to be able to talk to people privately as well as publicly to get any proper sense of things. You can't do that with a video strapped to your arse. Freedom to blow it in private before you come out with the bits you're getting right is a minimum condition of coming up with the goods."

In the first few months of the project the scientists and Jamie McCullough worked on the design and creation of the wetland. Jamie McCullough was not afraid to question the scientific input and had his own ideas of what a wetland should consist of. He was very demanding both of the scientists and Jozefa Rogocki, who was the 1½–2 days a week co-ordinator of the project. Often he was unwilling to indulge in the finer points of collaboration and he was not very interested in such aspects of Seen & Unseen as the schools work and community development which, he felt, were on the periphery of the project. He was passionately committed to the development of a wetland which would function effectively both aesthetically and scientifically, and applied himself tirelessly to the task of investigating its potential.

In April 1996, after six months of intensive work, Jamie McCullough decided to abandon the project as he felt that he could not resolve the work in the way that he wanted or bring it to a logical conclusion. Since the discovery of the old finings pond, the scientists had worked with two very distinct design constraints, the lack of height on the site and contamination of the soil. The scientific solutions and in particular the positioning of the clay embankments dictated the location of footpaths, seating and turning areas for wheelchairs.

Jamie McCullough had a dream of the wetland having pure water and fish and he was unable to reconcile this dream with the constraints of the design, both artistic and scientific. The next logical step for him was to "get out and work on real projects again" using some of the information that he had generated over this period to fuel future work. Sadly, Jamie McCullough has since died and we hope we are not doing a disservice to him in the way we have interpreted his reasons for pulling out of the project.

With hindsight, Jamie McCullough was probably not the right person to work on a collaborative project; his personality was not instinctively consensual and his view of how the wetland should be created led to disagreements, particularly with the scientists. Despite the disappointment in the way things had worked out, the group felt that there were some very positive aspects of his residency, some of which had a bearing on the rest of the project. He had challenged the scientists and in doing so encouraged members of the community not to accept at face value, without question, the advice and actions of experts.

The project was put on hold whilst the support group decided the best way to proceed. Some of the budget had been used up and therefore any replacement artist would have to be offered a residency over a shorter time span. A letter was sent to all the funders explaining that Jamie McCullough had resigned and that the Seen & Unseen were in the process of appointing another artist.

The search for a new artist

The post was advertised and three candidates were short-listed for the residency. Helen Smith, the artist eventually selected, had recently graduated from an MA course that she had undertaken after working mainly in residency, education work and on her own studio practice for eight years. This included a one-year post as artist in residence at the Louisa Leisure Centre in Derwentside. She had recently completed a voluntary collaborative project with artists and architects that focused on the urban environment and was based in Newcastle. She had also established a studio and gallery, the Waygood Gallery in Newcastle, with another five artists. In her application she described her practice as interactive. She explained:

"The process of collaboration is essential to how I make my own work. There is always a need to exchange ideas, methods and solutions in what I do."

The description of her proposed strategy for work was divided into two distinct sections, private work (the Unseen elements) and public work (the Seen). She anticipated that within the private aspect she would develop an understanding of the scientific process and the development and construction of the wetland. This she regarded as "the heart of the project" because:

"The wetland will surely need careful maintenance and enthusiasm from generations of residents. This is where I see the role of art in this project. To create a real ownership and pride in this solution there needs to be a genuine understanding of the issues involved in the creation of the problem, its solution and the environmental and political context."

She intended to create tiny site-specific digital images that would be placed at "points of receiving water in the homes and workplaces of people involved in, and supportive of the project … these images will form a network of individuals conscious of this project and the issues surrounding it."

The public aspect would document the project using video, photography and audio methods of collaboration to disseminate issues raised at Quaking Houses on an international scale.

Ian Jeffrey said of Helen Smith, "She had a more in tune, down to earth attitude, much more in line with how I thought someone would tackle it because she already had ideas about the need to talk to local people and so on. Some of the other candidates said similar things but she seemed to ring a bit more true…"

Paul Younger was also of a similar opinion. "I was in favour of Helen Smith, I thought that she interviewed very well… Of the three short-listed, she seemed the most normal, seemed the most straightforward. I didn't really care what kind of art she was into, as long as she would do art and as long as she was willing to do specific things."

It was agreed to appoint Helen Smith and she took up her post as artist in residence at Quaking Houses on 6 January 1997. At the inception of the project she agreed a timetable with Artists' Agency whereby she would have set up her studio and have done research on the gardens and wetlands by the end of January. She also agreed to find an environmental journalist with whom to work.

There is no doubt that Helen Smith had joined a team where her collaborative skills would be stretched. She started the project knowing the previous history, and with a further complication in that the project was well behind schedule and villagers were still awaiting the solution to the original problem of minewater pollution. Additionally, she had to fulfil the role of the artist on a much-reduced budget with equipment still needing to be bought.

6 Completing the project: The artist's perspective

By June 1997, it was envisaged that the wetland would be constructed and the ongoing collaboration with the villagers and engineers would be paying dividends. Helen Smith had decided to produce tiny images in response to her research and designs for the wetland. It was hoped that by September a publication would be produced detailing the method, approach and outcome of the collaborative process.

Understanding the science and scientists

From the outset of her residency, Helen Smith made it clear that she wished to be involved not only in the design of the wetland, but that she also wanted to work with other professionals to enhance the design. Having perhaps had their enthusiasm partly dampened by their experiences with Jamie McCullough, the scientists were wary of the artist's involvement and not as sensitive as they might have been to the new artist.

Adam Jarvis revealed that he had found it to difficult to ascertain where the artist was coming from and what Helen Smith actually wanted from him. He explained, "I get the impression that an artist starts with no idea of what they want and then wants to investigate every single option and then decide which of these options is most appropriate; but we scientists are very much more focused. We know that we have to do this… it's all very black and white."

Understanding the hierarchy in the university department was difficult for Helen Smith. She found it a way of working that was alien to her and had difficulty coming to terms with her role as an artist and her relationship in that role with Adam Jarvis. She countered this by focusing on a more one to one relationship with Paul Younger whom she felt appeared to be more understanding of where she was coming from.

Paul Younger responded on behalf of the scientists:

"I think it's one of the misconceptions from the artists' side that scientists are somehow not creative, not imaginative, that we somehow just follow an amoral process. That's not the way we work at all. It's extremely imaginative. I initiate my work to be creative and imaginative. Sure, there are backwaters of science that are extremely well funded that are just, 'well great, let's catalogue the next molecule'. It's boring, that's what we in the business derisively call butterfly collecting. The endless cataloguing of trivia. The reason that I like being a scientist-cum-engineer is precisely because you have to use your imagination big-style."

On reflection, Helen Smith realised that she was "throwing up ideas without any constraints whatsoever—design constraints or funding constraints. I think that I have become really aware that is what I was doing. I now have to find a way to balance out the core of the idea with all these constraints, which is what I am now getting involved with."

The members of QHET on the support group were impressed with Helen Smith's approach to understanding the scientific process and felt that the scientists could perhaps have made more of an effort to understand her artistic process. Terry Jeffrey summed this up:

"Helen Smith needs a lot of pats on the back because one of her first approaches was to try and actually understand the science. She started to read about wetlands and the science behind them, trying to understand by asking lots of questions and I don't think that the reverse is true."

In the interim between Jamie McCullough leaving and Helen Smith being appointed, the scientific process had undergone a rigorous investigation and experimentation. This resulted in a design for the wetland that could be built with minimum disturbance to the land. By the time that the wetland was built in August 1997, there was only one major change to the design. The scientists swapped the use of clay as a material for the construction of the embankments in favour of pulverised fuel ash, which sets hard when it comes in contact with water. There were three reasons for this decision:

- financial (the only cost was in transport);
- it is a highly alkaline material so it helps to reduce the acid in the water;
- it reflected the industrial legacy of the site and contributed to the ethos of sustainability by recycling.

Paul Younger was excited by the significance of this material, "it's ash from the burning of coal, so, it's symbolically appropriate to use it to deal with a problem to do with coal extraction."

Quaking Houses.
PHOTO: PAUL YOUNGER

The only material used in the construction of the wetland that was not a recycled waste material was limestone. Compost came from local stables, the sandstone from the adjacent waste transfer station, PFA from electricity power stations and the topsoil from local farms.

Paul Younger noted, "It's waste materials dealing with waste … it's almost like homeopathy."

Helen Smith's role as the artist

For Helen Smith, this was obviously a difficult time. Having entered the project halfway through completion, it was important to have her role acknowledged by members of the support group and for her to make a contribution to the final design elements of the wetland.

Ian Jeffrey from QHET felt that Helen Smith still had an important role to play even though major decisions about the construction of the wetland had been made. "It was no less an artistic job than if she had designed it herself because there was still plenty of work for her to do. We had identified many issues other than the construction of the wetland that the artist could be involved in during the time that Jamie McCullough was gone and we were artistless."

Helen Smith contributed to the final wetland design in several different ways and explored many different ideas during her period of research. The early work that she undertook with Adam Jarvis proved to be the most informative, and from this she came to realise that there were four potential areas where she could contribute to the design and therefore integrate with the scientific process. These areas were:

- the way in which the wetland is used, what kind of place it is;
- access and bridge design;
- design of the islands within the wetland;
- monitoring and therefore the future of the wetland.

Prior to any real involvement with the village, Helen Smith explored the idea of producing "a special place that is owned by the village. My work will be to research and share information about a broad range of private and public gardens from around the world, their formal qualities and their cultural significance. From this base, the villagers and I can identify what will work for us. Taking account of the practical constraints of the wetland, we will design the structure and planting of the garden in collaboration with the engineers. My starting points are Japanese gardens and the roles they play in the lives of the people who use them. They are places for contemplation, a place to rest at the end of a short walk from the home of the visitor. This seems relevant to the wetland outside of the village."

In addition, she had an interest in the use of sound; she was interested in developing the concept of hanging objects that are placed in the Japanese garden to catch the wind. At the wetland in Quaking Houses, she proposed that these could possibly be made by casting the metals (iron and aluminium) that are extracted from the wetland. These cast elements could also be embedded into the surface of seating or paving.

Once in post, Helen Smith decided to apply the concept of *access* both to the wetland as physical entity, and to the ideas embedded within it.

Collaboration with the community

By being more realistic in her approach and genuinely trying to fulfil the collaborative process in her position as artist in residence, Helen Smith worked hard at trying to involve all members of the community in the process despite the difficulties of the situation in which she found herself.

At the outset of the project Terry Jeffrey had introduced Helen Smith to the Village Hall Committee where she received a rather mixed reception.

Not all villagers had realised that Jamie McCullough was no longer the artist, and many villagers had their own idea of what an artist should be doing as part of the community project. Some members wanted Helen Smith to run the under-eight art club and others wished to charge her for use of the village hall. Terry Jeffrey commented, "I've never asked her really but I don't think that she was overly impressed by the Village Hall Committee. She didn't stay long, she just introduced herself."

Helen Smith had some frustrations. She was only engaged on a part-time basis and the portakabin that she was working from was not fully equipped with all the IT equipment which she herself had identified as necessary to develop her ideas for communication and access, and an offer of support in kind for this equipment necessitated a long wait.

Yet people wanted action.

Terry Jeffrey, knowing that Helen Smith wished to make small digital images, put together a list of people who would be willing to work with her and who would immediately welcome her into the village, "so that would get her started and whatever happened after that would be up to her."

Terry Jeffrey also arranged a meeting with Ossie Barret, one of the oldest residents in the village, who had written the history of the whole area and who had a collection of photographs of yesteryear. Helen Smith immediately saw the potential and intended to develop a digital archive that could be put on the world wide web and shared by local communities who had access to the Stanley Infonet. Julie Ross, the project evaluator, suggested that this could be linked to a web site at Glasgow School of Art that documented the actual wetland project using an interactive format. Sadly, Ossie Barret became ill and this part of the project was postponed. There was some talk of Lynne Conniss from Derwentside District Council's Stanley Infonet project taking on the web page design on behalf of Helen Smith. In the early stages of the project, Helen Smith refocused her ideas from that of creating digital videos to recording conversations with the local residents. She began her process by going on a walk around the proposed wetland site with children from the local youth club and recorded their impressions.

Camera Obscura held a two-day workshop with the local youth group, which encouraged the members to participate in an exhibition of their own exploring what was meant by Seen & Unseen. Within the workshop, three different groups explored the theme of Seen & Unseen and created a series of images, which told the story of pollution in their village. One surprising outcome came from the girls' group who wanted to produce a series of images on domestic violence, which they felt was a closer reflection of the Seen & Unseen in their lives. Alongside this event the groups also produced a video diary/documentary of the process and issues arising from it.

The audio soundscape was then played back to people in the village when Helen Smith opened up her portakabin studio to the village on the day of the youth club's exhibition. Members documented their own perceptions of Seen & Unseen at a weekend workshop using a giant pinhole camera and their own constructed stage and photographic sets. They also used video and audio recorders to interview residents around the village. This added to the participatory nature of the project by generating a number of different perspectives to the data collected. Links were also made with a number of different projects including site visits from school and college groups, and a link with a school in Norway.

Derwentside District Council gave the village a computer with internet links as part of the Stanley Infonet project, which sought to provide a range of community organisations, schools and colleges in the Stanley area with computers and links to the internet. The council also offered six training places to people in the village to learn computer and internet skills. It was decided that Jozefa Rogocki, Helen Smith, Terry Jeffrey and three recent school leavers from the village should attend.

Later in the project, an opportunity arose for community groups to undertake radio training as part of Visions of Utopia, another Artists' Agency project that was being managed by their co-director, Esther Salamon. Radio Utopia was directed by the Communications Department at the University of Sunderland. Helen Smith, Lucy Milton and Jozefa Rogocki responded to their call for interested community projects and devised a project whereby Helen Smith, members of the youth group, a youth leader and Terry Jeffrey received training in interviewing and radio broadcasting techniques. This resulted in two live radio broadcasts from the University of Sunderland's temporary radio station during the Visions of Utopia event. Helen Smith based her proposal around the idea that the residents in the village would be able to listen to the broadcasts and thus engage in a wider debate about the wetland. Sadly this was not possible as Radio Utopia could only be heard within the Sunderland radius and Quaking Houses fell outside the reception area.

In order to allow residents of the village to take part in Helen Smith's project, a broadcast quality tape recorder and editing software for the computer was purchased. This gave the youth group the opportunity to pass on their training to other people in the village, who could then make their own audio programmes for radio or the internet. Not only was this a good practical idea for community involvement and skill development, but sound was something that Helen Smith was particularly interested in as part of her work as an artist. She also felt that her radio work was a way of establishing more meaningful relationships within the community both between the residents and with herself. In terms of the wetland and establishing a wider understanding of the place, Helen Smith noted that "it's stirring up all sorts of discussion. I feel as though I am creating a space in which there is a dialogue between people that I can enter into, in which they can listen to their stories and tell me their stories, which then informs me and what I am doing in the site."

After the Radio Utopia event Helen Smith said that "one aspect of what I chose to do in terms of working with people in the village, in relation to the wetland, was just to pick up on this thing of live radio where the people could debate the issues. It seemed to me that to arrive on a project and to ask young people, in particular, to get immediately involved in a field in the middle of nowhere, you were on a hiding to nothing. It didn't even interest me very much either at that time, it was just a field. I felt that the significant thing about this project is the issues involved in it. The history of environmental action within the village was what was interesting. The role of radio—the potential of radio and the journalism strings within that—seemed to me to be a highly interesting and appropriate one which could grow with the project once those skills were in place. If I worked with one group, they would inevitably then get around all the different players in the project. They could then produce something which could be shared with a larger audience and also bring themselves in direct contact with the subject and working knowledge of the project. It worked in some ways and it some ways it didn't work. The project was there, we gave training, we produced two live programmes, which were then broadcast, Terry Jeffrey was involved and I think that it was a tremendous experience for them. That hasn't stopped either. As an extension of the project, I have started doing some research in the radio department at Sunderland University and that has funding to continue into a women's radio project. So that's all in place too and continuing."

Helen Smith kept her own personal log of the project. Shared documentation of the project was undertaken by a number of different people including the scientists, the artist/evaluator and photographers. Ian Jeffrey, who received two days of training in the use of a digital video camera, documented the construction of the wetland, together with members of the youth club. Yet despite all these achievements, Ian Jeffrey felt that neither the youth club members nor Helen Smith had appreciated the time constraints on each other. He summed up the feelings:

"She [Helen Smith] was in a bit if a tizzy about putting her stuff on because it was seen to be a bit of a rush job in the end for her. They [the youth club] felt that she had not taken a blind bit of notice of their exhibition because she had come in, grabbed something and rushed out again to her portakabin and when her bit was done she went home."

In essence neither appreciated the other's problems. The youth club leaders did not explain the constraints Helen Smith was working under and she did not go out of her way to explain to the children that she had other commitments beside Quaking Houses.

Helen Smith explained her feelings thus: "I found it very difficult being in the village. I had a very isolated year in that respect. I've been working on it three days a week and it's been tough… I've got it in the neck from both sides, from the people on the site and from people in the village, because I was constantly somewhere else. People didn't know that I was actually tearing around in a car trying to do sixteen different jobs, like everyone else in this project. I couldn't have worked any harder, and I know that."

Another major achievement and a direct result of Helen Smith's involvement was that one school leaver in the village decided to take up art seriously and enrolled for an art course at a college of further education.

The wetland – artistic input to the design

Helen Smith was of the opinion that the scientists' design was not really welcoming people into the heart of the wetland. It merely provided a circular walking route with viewing points into the wetland. Jozefa Rogocki was also very concerned about the reasons for wanting to bring people down onto the site and questioned whether the support group was trying to create a haven for wildlife or a picnic area. In her opinion it was apparent that it had been decided that it would be a good thing to give people access to the site— yet no one had bothered to ask why.

Terry Jeffrey summed up the feelings of the residents, "I think that the point is that you shouldn't over complicate it, because it's not that type of

Helen Smith (left) helping to build the wetland site.
PHOTO: PAUL NUGENT

site, is it? But you've got to provide access to it, because all human beings have a right to visit it."

Helen Smith felt it important to engage a landscape architect to work with her because they had the expertise to look at all the usages especially from the perspective of the people that lived there. Adam Jarvis was not very happy about using a landscape architect and thought that the additional costs of employing one were unnecessary. He pointed out that basically they were building a pond, which would not have much land at either side. He felt that any paths were predetermined by the layout of the site. Helen Smith felt that there were also the islands in the middle of the pond and the planting to consider. She said, "If all we had to do was our own basic wetland then we would all come up with really different results. There are choices to be made and I just believe that we should look at this book that I am reading: you will see the history of landscape and landscape architecture and the way that we are affecting the environment with it."

Terry Jeffrey felt that if she could do the work within the budget then it would be all right with the residents. It was agreed that Helen Smith should collaborate with a landscape architect and that her collaboration could then inform her approach and contribution to working with Adam Jarvis.

Helen Smith envisaged that a landscape architect would act as a design consultant for her on drawings, materials and costing. She felt they would be "putting the practical element to what I have come up with on paper. Then there would have to be contact with Adam … to tie in with the construction of the wetland itself."

Working with landscape architects from Groundwork, Helen Smith took Terry Jeffrey's advice and looked at the leaflet on local walks designed by the Footpaths Group. Together, they considered how the area was being used by ramblers and—in consultation with QHET and an advisor on access for people with disabilities—Helen Smith finally proposed a timber walkway design (incorporating both the bridge and jetty) that would take people right across the wetland, entering along with the polluted water and leaving with the clean water. In keeping with Helen Smith's concept for the design of the wetland, form follows function.

John Knapton, a structural engineer from the University of Newcastle, joined the team as a consultant to advise on the construction of the final solution. Helen Smith also collaborated with Groundwork's landscape architects who took over technical responsibility for the footpaths, the gradient of the slope for wheelchair access, and linking the access route through the site with the existing footpaths.

Rather than an open channel, lined with a material that would collect and show the pollutants on entry to the wetland, Helen Smith intended to incorporate two glass panels into the actual walkway. People could look down and see the polluted water flowing in under their feet at the beginning of the walk, and clean water at the end of it. She also wished to incorporate minerals embedded in cast glass into the handrails and to commission a glass designer to sandblast a design onto the glass panel. In September 1997, she took some of the youth group to Hartley Wood's glass factory in Sunderland where the glass artists suspended iron ochre (collected from another polluted mine site in the region) in the glass. This was an experiment and the glass experts were of the opinion that the chemical reaction between the irons and the glass would over a period of time cause the glass to shatter. Helen Smith's suggestion at the support group meeting was to use sandblasted images that depicted the scientific formulae from most pollution to least pollution. Following further research into glass processes the glass was ordered. Due to the size, it was manufactured and toughened in London before being sent up to Newcastle where Helen Smith intended to have it sandblasted. By November, when the wetland was launched, this had not been done. The glass section at the influent was installed, minus the sandblasted images, and the other glass panel at the point of exit for the clean water was missing. This was due to be installed at a later date. But sadly, by the following month the original panel had been vandalised and the glass broken. Although delighted with the boardwalk, residents became concerned that its design meant that the construction could be easily dismantled and the timber used for garden sheds or even pigeon lofts!

It was anticipated that the first phase of the jetty and the wetland would be completed by July, but bad weather postponed this until September. By mid-September the earth works were complete but the jetty was not in place. Delays had been caused by the added structural and construction problems caused by the extension of the jetty. Aside from these problems, Paul Younger was responsible for ensuring that the original planning application was updated and the extension approved.

It had been agreed to launch the wetland site on 23 November and the delays and worry that the work might not be completed on time led to an exchange between Artists' Agency and the scientists.

Paul Younger was particularly concerned that if the flooding of the wetland could not commence sooner, the absence of water for a lengthy period of time might have an adverse affect on the bacteria in the compost necessary for the cleansing of the water. Out of sheer frustration Paul Younger wrote to Lucy Milton on 23 October outlining his concerns and concluding that as a result of these they would go ahead and flood the wetland in time for the launch even if the walkway was still incomplete. In the end, this was not necessary as Helen Smith ensured that the basic structure of the walkway was ready for the launch.

Design of the islands within the wetland

Because of the complications in understanding how the wetland would function, Helen Smith sought the help of Mike Riley, a hydrologist and mathematical modeller in Paul Younger's group at Newcastle University. He showed her how the design of islands in the wetland, their shape, gradients, and the materials used in their construction would influence the working of the wetland by affecting water flows. Helen Smith wished to use the shaping of these islands as part of her aesthetic vision of the project, and felt also by involving herself in their design that she would be contributing to the functional purpose of the design.

Adam Jarvis had on several occasions made it clear that he felt there was no need for the artist to bother with the scientific detail and Helen Smith was equally sure that he was not particularly interested in the art. To be fair, part of the reason for the lack of collaboration at this point was no doubt due to the fact that they were all under pressure to complete the wetland as soon as possible.

Helen Smith felt frustrated and furious with the perceived reluctance to allow her to go beyond the aesthetic and consider function. She pointed out that when she had started on the project, wetlands were something new to her, but through her research she had developed certain knowledge and a point of view, which ought to be considered. If Adam Jarvis was not prepared to listen to her, she needed to talk to someone who would. She said: "I need to talk to someone else about this. I think it's an area that the previous artist could not get past and I think it's significant that we did get past it."

Paul Younger felt that Helen Smith had not got her facts right and that, sympathetic though he was to her desire to understand the science, she could not realistically become an expert in the complexity of wetland science overnight. He felt that Adam Jarvis was not being given the respect he deserved as a professional. Paul Younger also felt that although the islands appeared to the visitor as islands, they were in fact barely submerged spits whose function was to divert the water, the science of which had been tried and tested in the pilot wetland.

Helen Smith's original intention was to learn how to use the computer software to produce her own designs. However, it transpired that she did not have the background knowledge or the necessary time to understand both the complex mathematics and related software. In the end, Mike Riley helped present solutions to Helen Smith and Paul Younger and these were incorporated into the wetland design as the three islands and baffles described earlier.

Helen Smith was grateful to Mike Riley and appreciative of the time he gave her. She admitted that a degree in hydrology would have been necessary to understand the principles behind the island designs but she pointed out it was Mike Riley, not Adam Jarvis, who taught her about water movement. She said, "We aren't trying to slow the water down as I was previously describing but we are trying to spread it out so that it moves across as much

View of wetland and jetty.
PHOTO: PAUL NUGENT

of the compost as possible. From an artist's point of view that's a really important piece of knowledge. Isn't it?"

Adam Jarvis maintains that he had previously explained the importance of the above to Helen Smith, and it can only be assumed that because of the present difficulties of the collaborative process, Helen Smith preferred to communicate with Mike Riley. Paul Younger pointed out that although the work that Mike Riley contributed to the design was useful, it was not as definitive as Helen Smith believed. The work that Adam Jarvis contributed was the essence of the functioning wetland.

The scientists had much more fixed views about Helen Smith's role and the design of the islands. Paul Younger was more interested in the potential of the islands as ecosystems. He saw potential for sculptures on the island— similar to those that he seen in various examples given by Jozefa Rogocki at the start of the project. He recalled that Helen Smith had suggested sculptures but this had been put aside as she followed her interpretation of 'form follows function'. However, he wondered if this wasn't Helen Smith's style and if so, why Artists' Agency had not used any of their funds for additional commissions for this purpose.

Monitoring and the future of the wetland

Funding, however, had been awarded by Agenda 21 for the long-term monitoring of the wetland by residents of Quaking Houses who were to be trained by the scientists from the University of Newcastle. No provision for long-term monitoring was available to the artist from this budget but Helen Smith proposed a series of sculptural monitoring stations that would have to be funded from the main budget or with additional funds. Helen Smith said that to date, "I've come up with a sculptural intervention which Paul Younger and I are working on to do with the monitoring and making that accessible to people. I don't have much to say about that but Paul Younger has some good ideas about what sort of information the department would like. It would also be useful on an ongoing basis if we can involve QHET in the upkeep of some things and in the transferral of that information. So it's my job to come up with an image and a sort of design proposal as to how that would work."

Helen Smith was also interested in using voice as signage or as monitoring devices, whereby the final piece would take the form of an installation of audio sound boxes at the wetland site. These would be operated by push button, allowing anyone visiting the site to listen to the results of monitoring that would have been pre-recorded by members of QHET and updated on a continuous basis. The 'sculptural listening posts' as they were called attracted funding just as Helen Smith was leaving the project, so their development was taken on by Steve Smith at Radio Utopia and designers from the University of Northumbria.

As with all projects that require constant fundraising, rarely do the funds accommodate every aspect of the project. This is inevitably a cause of frustration and even more so when engaging in a project that relies on several different participatory elements coming together. The community understandably wanted to see certain parts of the project take precedence, likewise the artists and scientists. To the members of QHET, planting was a priority and it is worth examining how this issue relates to the collaborative process.

Planting within the wetland became a future issue because by the time that the wetland was actually built and functioning, there was no money left in the budget and it was mid-winter, not the true planting season. Adam Jarvis had thoroughly researched the area but did not have the funds to put theory into practice. There had also been visits to the site from wetlands experts at Newcastle University who had offered advice to members of QHET. Planting within wetland had shifted to an aesthetic and ecological issue, rather than one of function, because the final wetland design solution no longer required the use of plants to take up any minerals and clean the water. As Paul Younger pointed out, "The wetlands don't need the plants in the short term for the treatment to work. We would have been happy for all plants to colonise naturally in any case."

But the residents at Quaking Houses felt differently. Alan McCrea from QHET had undertaken a lot of reading on planting and was keen to get involved. By March 1998, the planting season upon them, the members of QHET were becoming quite frustrated and angry with the whole issue, as no one seemed to want to take responsibility for the planting. The simple truth was that the project did not have the money available and it became an issue whether it had in fact ever been budgeted for.

Terry Jeffrey on behalf of the QHET said, "We've got what we wanted: clean water. If you look at the water that is now pouring out of the wetland, down the burn, it is clear. So as far as that goes we are happy, but we did think that we were going to have quite an attractive thing for people to come and look at. In our view we are not quite finished yet, and one thing that we must

Adam Jarvis examining water samples at the wetland site.
PHOTO: PAUL NUGENT

do and we must do now is the planting—because mother nature will wait for no one."

Part of this delay was due to financial constraints. As Lucy Milton explained, "We have gone way, way over budget to the point that Artists' Agency has actually been paying for things themselves and we are using up other budgets that were allocated for other things on this project."

Derwentside District Council, who at one stage in the project had offered to provide fencing and planting along a section of the site, were now no longer able to fund this because of annual budget restrictions due to the time delays.

The scientists and members of Artists' Agency felt that planting had always been seen as desirable but had not actually been budgeted for in the project: Lucy Milton stated that she would need to seek further funds.

Terry Jeffrey felt that he didn't have that long, nor did the other members of the QHET who were willing to help with the planting, because they were all in or approaching their seventies. They had begun the process of cleaning the water and they wanted to see the area finished so that "we can say that this phase of the wetland is now completed and that we are going to monitor it. We need to have dates and we need to have times."

In spring 1999, artist Lee Dalby started to make living willow sculptures to complement the wetland. Members of the youth club have planted marsh marigolds and other bog loving plants along the banks. But Terry Jeffrey's prediction that he would be unable to help with the planting because of his age was tragically true. He died not long after he had accepted awards for the wetland on behalf of Seen & Unseen. Chas Brooks, Alan McCrea, Tommy Cole and Maureen Wilson from Quaking Houses are now caring for the wetland, spending time planting saplings, tidying up and carrying out minor repairs.

The launch, the exhibition, and the winning of awards

Following completion of the wetland, it was agreed to host a launch in November. After much discussion, a lantern procession was arranged, Jozefa Rogocki having organised and participated in a schools workshop to make the lanterns.

Residents, the scientists, artists and invited guests strolled the half mile to the wetland site by the light of the candles and the beat of a local drumming group, whereupon Terry Jeffrey proudly declared the wetland officially operational. There followed a firework display and a dance back at the Quaking Houses village hall. Terry Jeffrey, remembered this as one of the most moving events of the whole project.

The achievements of Seen & Unseen were recognised by the two major awards given to them for innovative and collaborative practice. The Henry Ford 1998 European Conservation Awards gave the project £5000 as a category winner and allowed the project to go forward as the UK contender representing Britain at the European finals against thirty-four other countries. The wetland also received a commendation in the BURA Charitable Awards 1998. Chas Brookes from the QHET accepted a certificate of Best Practice at an award ceremony at the House of Lords on 8 May 1998.

The opening evening on 12 November was truly a celebration of a collaborative project. Past difficulties put behind them, all partners could relax and view with pride their achievements over the past years. The atmosphere was one of acknowledgement for each others' contributions.

From 13 November 1988 to 8 January 1999 the exhibition, Seen & Unseen, ran at the Hancock Museum in Newcastle-upon-Tyne. Helen Smith had found a wonderful old writing desk to show her interactive interpretation of Seen & Unseen (see Plate 14). Hidden drawers played recordings when opened and small digital images could be seen. The newly created website was on display taking visitors on a virtual reality tour of the wetland. Further planned events include small touring displays and a conference.

The listening posts have finally been agreed and will be in place in the near future. Visitors will have an opportunity to hear the story of the wetland. Helen Smith conducted many recorded conversations with all those engaged in the construction of the wetland and these will form part of the tapes for the listening posts.

Terry's dream

Terry Jeffrey was one of the inspirational minds behind the planning and building of the wetland. It was his dream, his vision, and improving the quality of life for future generations in Quaking Houses was his guiding force. Without his determination, sheer hard graft and persuasive powers, this project might have fallen at the first hurdle. He believed in the concept of collaboration and interdisciplinary projects; his reference to C. P. Snow's *The Two Cultures* where he highlighted the need for art and science to collaborate was pivotal in pointing out to other members of QHET the potential of Seen & Unseen. He was well read, passionate about the future but also determined that people should not forget the hard life miners endured. We remember him taking us around the village, pointing out the old mine shafts and telling us the story of a day in the life of a miner. He proudly showed us the village green and the landscaping that had transformed the village, but never forgot where its roots lay.

We are all proud of Terry Jeffrey and long may he be remembered in Quaking Houses.

7

The project as an actuality

"Is there a distinction between collaborative vision and collaborative solution?"

Following the completion of the wetland in March 1998, David Butler (who edits *Artists Newsletter*) posed the above question. The simple answer is 'yes', visions surely are much easier than the process, but the question is really a highly complex one and the answer is dependent upon many factors. Turning visions into reality involves the complexities that the human being brings into the process. Interdisciplinary collaboration requires the participants to forgo the luxury of being too precious about their own discipline and develop an understanding of another point of view. More than anything, it requires a freethinking spirit prepared to view the project as a holistic enterprise.

When designing a new project, especially an interdisciplinary project involving arts, sciences and community, it would be unrealistic to think that it would be free from pitfalls.

The wetland project began when the QHET approached the authorities to ask for help in cleaning up the Stanley Burn. The idea for a collaborative process came later. It is very difficult to assess whether the community would have felt satisfied with just a clean stream free from pollution without the artist involvement. There is no doubt that without artistic input there would not have been an opportunity for the residents to learn new skills in the field of sound and print media. Whether there would have been a public space that is engaging and practical is also a moot point. There certainly would not have been a boardwalk, exhibitions and the opportunity to disseminate information in the way that has been made possible with funding obtained mainly by the efforts of Lucy Milton from Artists' Agency.

Throughout the project, there has been a tendency from both the arts and the scientific community to want to take the lead role, yet the truth is that each discipline has had to listen and respect the other's viewpoint. This has been shown in previous chapters outlining the history and progress of the project.

So is there a distinction between collaborative vision and collaborative solutions?

The answer depends on who has the collaborative vision, who proposes the solution and the context in which it is required. To an extent, it is also determined by the requirements of the funders and whether their input is active or passive. This in turn decides which agency manages the purse strings and whether the agent that holds the money has the final say when consensual decision making becomes cumbersome. We have noted that there has been a sense of disappointment between some of the players regarding the division of the finances. We have seen throughout this project that ideas that could have added to the collaborative vision have been dropped because of lack of finance. This means that both the vision and the solution are susceptible to change at short notice. Another important point is the selection of the personnel involved and their personal desire and motivations for the need to collaborate.

Would the project have been significantly different if the dynamics of gender had been different? The key scientists were all men, the QHET was predominantly male, whilst the artistic input was predominantly female. When Helen Smith was appointed in place of Jamie McCullough, the artistic input became totally female. Therefore the stereotype of men studying science and women the arts was reinforced throughout the project.

The project also lacked a full-time permanent co-ordinator whose remit should have been to take an overview of the process as a collaborative enterprise. The role of part-time project co-ordinator was taken by Jozefa Rogocki, an artist from Darlington, originally for 3 days a week, later reduced because of the extended time-scale of the project to 1½–2 days a week. She had the impossible task of trying to be all things to all people. Reflecting back on the project, and a developmental budget in the region of £150,000, it is unfortunate that because of constraints on the use of the various sums available a full-time co-ordinator could not be appointed, independent of both the arts and sciences, to manage Seen & Unseen. We feel that should other projects emulate the achievements of Seen & Unseen, it is important that an independent project co-ordinator is appointed who reports directly to a support group type of structure.

Wetland looking towards waste transfer station.
Photo: Artists' Agency

We think it true to conclude that at the beginning of the process, members of the support group came to the meetings with an unrealistic view of how the project would work. Considering that all the players were entering new territory—the theory was laid down on paper but the putting into practice of this new vision was a new experience—it is not surprising that each member of the support group had their own idealistic viewpoint based around what they each felt collaboration meant. Sometimes these views coincided but often the differences were not communicated in a way that led to a constructive resolving of the dispute. This is where we believe an independent project coordinator would have had a role to play.

Although Jozefa Rogocki took on the role on a part-time basis, as an artist in her own right she had other projects she was working on. She certainly more than fulfilled her contractual obligations, often working far longer than the one day a week she was called on to do. However, Seen & Unseen demanded more than that. This meant that, in essence, the role of overseeing the project fell on the shoulders of Lucy Milton who, because of her role as Co-Director of the Artists Agency, was obviously seen by others as a champion for the artist. However, Lucy Milton's personal belief in the collaborative nature of the project and her skill in the art of diplomacy allowed her to take a role that encouraged a consensual viewpoint. We conclude that had Lucy Milton not been in this role, the problems that arose from the lack of collaboration would have been greatly magnified.

Lucy Milton genuinely believed in the collaborative process and illustrates the success of that process in a number of areas.

She was of the opinion that "much of Helen's work is unseen; I think that's the thrill of the project—that it's difficult to say, 'that's the artist's bit'. The whole point of collaboration is that the work is integrated into the whole, and cannot be separately identified. I hope that we will all be able to pat ourselves on the back about this holistic approach—it's certainly of interest in the arts world, and I hope it could also be of interest to the scientific community."

Lucy Milton acknowledged that perhaps the overall collaboration between all participants in the project had not been as successful as she would have wished, but she believed that the outcome of the project showed that this holistic approach could mark new ways of working.

The scientists had felt at times that they had been submerged into the background and not given due credit for their technical input into the project. Had the engineering part of the project been enhanced by art or indeed was there a perception of engineering as art?

There is no doubt that both Paul Younger and Adam Jarvis did feel at times that they had not been given due credit, either for their technical input or for their status as full collaborative partners. They felt there was a perception that they were seen as technicians brought in to make the water engineering work, but that this was a basic necessity that was somewhat divorced from the more important 'cultural' aspects of Seen & Unseen. This perception was intensified by an awareness of the financial shareout between the 'scientific', 'artistic' and 'community' partners. It also begs the question of whether the engineering side of the project had been enhanced by the 'art', visually or technically; and indeed whether what Paul Younger and Adam Jarvis were able to bring to the project could also be counted as 'art'. Certainly, while the chemistry of the processes taking place in the wetland is not easily understood by lay observers, what is clear is that these complex processes are brought about by relatively simple means. In the simplicity of this engineering solution, in the sheer neatness of the solutions adopted, in the appropriateness of the use of organic and recycled materials to clean the polluted water, and in the form of wetland adopted (which is perfectly good engineering for all that it does not require oblong concrete tanks), there is a beauty which, it could be strongly argued, makes the engineering itself part of the artistic input. This was consciously reflected by Helen Smith in the layout of the boardwalk winding across and past the wetland as the water flows through and, particularly, in the proposals for glass panels in the boardwalk at the start and finish of the process, tying the artist's and the bystander's views very directly to the engineering problems and solutions: dirty in, clean out.

It is also clear that at the beginning of the project, Paul Younger was excited at the inclusion of an artist. His remarks on the role of Artists' Agency and the creativity it could bring to the project were summarised by the statement he made after seeing the London-based arts group PLATFORM and reading the brief on the water project prepared by Lucy Milton.

Paul Younger said at the outset of the project:

"The whole idea of the Artists' Agency water project is that it can help to illuminate or illustrate these [hidden] things. As scientists we use a lot of visual things but usually a lot of creativity doesn't go into it. We use standardised forms."

Adam Jarvis, the scientific researcher, was no doubt a little less enthusiastic about the role the artists had played in the creation of the wetland and the extent to which collaboration had worked. Like the artist Helen Smith, his view was from the point of dealing with the collaboration on a day-to-day basis.

He summed up his feeling as follows:

"It's a great idea, collaboration, but there have to be very tight limits on that collaboration because it just would not work. I haven't tried to get involved in the artistic side of things because I simply don't know about them … and I don't see the point because we have an artist here."

This viewpoint must have been reinforced by his having to work with two artists, both of whom had different perspectives on what the Seen & Unseen element was. No doubt his frustrations on a day-to-day basis were then exacerbated by what he saw as the repetitive element of his work with the artists. He also was in place before Artists' Agency were involved and had his own ideas of what the wetland should look like which, at times, must have left him feeling somewhat disenfranchised in the whole process. It may have been that he felt his primary role was somewhat usurped by the increasingly large project with which he became involved.

Helen Smith concluded that collaboration very much depended on the individual and their own professional interests and desires:

"One of the key things that I think in all of this, is that individuals regardless of their profession actually need to choose to do it [collaboration] and be interested in that kind of progressive way of working… There needs to be that sort of personal energy."

She felt that in her experience, in most collaborations, it is the arts people who usually attempt to bridge the gap by moving into someone else's area and not always with an equality of movement from the other area. This of course is in direct contrast to the way in which the scientists had felt the project had been managed and possibly gives an insight into why Helen Smith was determined to involve herself in the scientific process.

Terry Jeffrey from the Quaking House Environmental Trust put it thus:

"It's impossible to credit one individual or one mind set for the total project. The total product is as a result of the collaboration, or should be."

Following the construction of the wetland, talk naturally turned to how to capitalise on the project following its success in winning awards. However, some participants in the support group felt that there was more to be done on the ground before reaching out to a wider audience.

The residents of Quaking Houses were delighted with the construction of the wetland but they were deeply disappointed that other aspects of the project, such as the planting, had not been completed. There was a general feeling that their wishes as a community were being ignored as the project moved forward—in their opinion, too quickly—into the national and international arena.

Their opinions were put forward by Terry Jeffery:

"You have succeeded to the extent where you have made the water clear. You've done that but you haven't completed the wetland and so to build on that is quite weak."

The argument for moving the project forward was not only the enormous cost of the exercise but the moral duty to share information with the outside world. Not only could Seen & Unseen, as a concept, benefit communities in the United Kingdom, but the relatively low-tech solutions to water pollution could help others, especially in the less developed countries. During the course of the project, Helen Smith had made contact with Penny Kemp, an environmental journalist who, on hearing about the project, was excited at the prospect of disseminating the information to a wider audience. Helen Smith told Julie Ross, the artist evaluator:

"What she thought was currently interesting out there in the world was the art and science collaboration on solving problems, and also the example the project makes to the power of the individual to actually get things going and just to keep on track. It [the project] has the potential to be of enormous value to communities both at home and abroad. However, unless publicity on a wide scale is actively sought, the outreach work required could easily be neglected and hence a valuable part of the project will fail."

She was of the opinion that what at first appeared to be a very simple project was in actuality "a surprisingly complicated project working on a number of different levels and reaching out to several audiences."

At the time, these views were academic because there was not the funding to put into action the ideas that Penny Kemp had put forward.

Not all members of the support group felt that extending the project was necessarily a good idea, much less disseminating information to a wider audience. Some felt there was pressure to publish virtually as soon as the project was complete but both Paul Younger and Julie Ross felt that this did not allow time for a proper analysis, given the duration of the project (four years instead of two) and its increasing complexity.

As Paul Younger puts it: "It's the first wetland that Adam [Jarvis] and I have ever built … it's a big deal for us as well."

Adam Jarvis felt he still had much to do in relation to the evaluation of the Quaking Houses wetland, but also felt that the results so far were of value to others:

"The scientists now have a research resource, in the form of the wetland, which from my point of view is the most important thing. We are now legally obliged to make sure that it continues to clean water and monitoring has been taking place on a weekly basis. This has resulted in a large and growing data-base. Further research is going on into what is actually happening in the wetland as the water is being cleaned. The results of the research are being disseminated in scientific research journals where the emphasis tends to be on the processes going on within the wetland. There may be other articles where the emphasis is on encouraging communities to become involved in projects. In the USA communities and volunteers are now doing similar work without the assistance of the universities and they have become very knowledgeable in this area. The scientists from Newcastle University would be very keen to promote this way of working."

Peter Woodward, from Shell Better Britain, one of the funding organisa-tions and a regular attender at the support group meetings, felt that there was an impatience to push the lessons out while a complex range of issues still had not been resolved. The challenge, as he saw it, was in completing the wetland to a stage where one could actually walk to the site and see the whole scheme, including planting, as a functioning whole. In his opinion the community needed to feel good about the results:

"There is still twice if not three times as long to go because of the complexity. I keep having to reinforce that the complexity does not imply criticism, it implies that you have landed on something that is bigger than all of us and one that has to be completed… The triumph is that clean water has been achieved and we mustn't lose sight of that, but also the hope was that the community feel somehow engaged, strengthened, proud—but until it looks great … the community hasn't yet got that."

Peter Woodward's remarks should be taken in the context of the aims of the Shell Better Britain campaign, which gives funds to help improve the community and allow the community to be involved in improving their own environment. Shell Better Britain would obviously wish to have a visual outcome that reflected well on their publicity programme. Discussions had already taken place about the role of Shell and whether artists should accept funding for an environmental project from multinational organisations that are seen by some to have poor environmental records.

Helen Smith the artist was keen to take the work on exhibition and reach the wider audience:

"I've been designing the exhibition or the road show to go to other arts venues and I think that we aren't being ambitious enough in relation to it. The sense of just not having the context for it is really strange here. It's too interesting and it's too new for what's actually going on artistically here."

Lucy Milton felt that Helen Smith's contribution to Seen & Unseen was, without doubt, central to the success of the project. She said:

"I think Helen has been amazing in the sense of the scope and range of the work that she has done for this project. From the very conceptual idea of taking interest from the villagers in terms of sound and finding functional ways of dealing with that—in terms of helping to create a functional facility that can be used in the longer term ... broadcasts ... collaborating on works about the past, present and future of the village ... incorporating sound into the sculptural monitoring stations ... it's just so amazingly imaginative and wonderful. It's taking the idea of an artist as somebody who actually conceives; what they create is part of everything. It's just so totally integrated into all sorts of social, political and personal dynamics... In terms of the wetland she hasn't tried to impose, she hasn't tried to create a physical work of art *per se*. The fact that she is looking very sensitively at how, through dialogue, she can have, perhaps in some senses, minuscule changes that are very important. The fact that she suggested that the water tank could be placed under the jetty instead of another location, and a glass piece over it, revealing the processes from the polluted to the clear, showing the symbolic nature of the whole regenerative process. That is so simple and yet so structural and so much part of the whole..."

Jozefa Rogocki thought that the sciences were often too eager to claim that the arts were taking from the sciences and advancing on the basis of research that they had already done. Whereas she believed that the opposite had actually occurred during the project. She commented:

"The arts is the cement, the bridge ... in this project and others... The walkway, the jetty is in a sense a metaphor for their role within this project. The arts provided access for the community into this kind of work, it took the initiative into work with education... I think that's maybe why the arts side more than the science side talk about this new role because we find in our own specialism of the arts that this type of work is perceived as being as credible as an arts form because one chooses the form or the process that is appropriate to the place you are working in and to the people that you are working with. It's a responsive way of working. It involves creating bridges and it involves participation."

Peter Woodward, from Shell Better Britain, picked up on the differing views of the way the collaborative process had worked, and believed that a tighter collaboration between the support group had not happened because, despite the best efforts of Artists' Agency, there had not been anyone involved who had the skill or specialist competence to co-ordinate such a complex project.

He summarised his thoughts on the issue of project management:

"What has happened in reality is that the simple vision [that we had at the outset] has become extraordinarily complicated but rather than taking stock and stopping and saying that this has become a two million pound, five year project and getting a full-time co-ordinator with these skills to help it, people have valiantly and wonderfully put more in the boiler to keep it going... It's been a fumbling through approach rather than saying that this is a more significant set of problems than we anticipated... The stepping back from and reframing of the project hasn't happened."

Lucy Milton had in fact made considerable, but ultimately unsuccessful, efforts in approaching major potential funders with a view to raising sufficient funds to allow, among other things, a full-time project manager for three years. But, she said, "because the project was so new, and there was nothing on the ground at the time to show for it, and no comparable examples, no-one wanted to know. So rather than wait for a hypothetical pot of gold, we proceeded with what we had."

Jozefa Rogocki agreed with Woodward but highlighted that Artists' Agency had always been aware of the breadth of the project and that it had originally been much wider than just the development of the wetland. In her opinion one of the fundamental problems had been potential funders' inability to grasp the complexity and breadth of the project and come forward with the money for what Peter Woodward had just identified.

She noted that:

"At the time Jamie [McCullough] resigned we had to confront a lot of these problems, Jamie raised lots of issues but we could not look them in the face. We had to continue. There were certain things that we had to tie ourselves to because there was no way that we were going to raise the money to support the things that we actually knew that we needed... I think that this is a symptom of the project where people are trying to achieve something that they know is more ambitious, of a larger scale. We knew the difficulties but because of a lack of funding the recognition of those roles could not be highlighted."

Lucy Milton had experienced the problem of obtaining funding commensurate with what they actually wanted to do.

She described the process as "trying to break the project down into bite-sized chunks where you have a range of overlapping funding interests. Securing funders didn't just happen, it involved going out, finding out, ringing them up, talking to them, individualising an application that made sense to them, researching the different kinds of funding that were offered within a particular company or trust, doing an application, inviting them to visit, taking them around the proposed wetland site."

Despite spending far longer on fundraising for Seen & Unseen than for any other project with which she had been involved, Lucy Milton was left with the dilemma: either doing things on piecemeal funding and with everyone struggling to achieve the goal, or not doing anything at all. She described her role as similar to that of a missionary, salesperson or advocate trying every avenue and making hundreds of applications until she could piece together enough funds to make some headway with the project. The funding received was usually for a specific part of the project and freedom to move that money around from one use to another, however beneficial that may have been, was limited.

Many of the unsuccessful applications were put down to the innovative nature of the project. There were no precedents to refer to and so potential funders were being asked to make a leap of the imagination. In order to overcome this problem, Lucy Milton looked at the types of people that specific funders had given money to in the past and she also tried to make the applications for funding as conventional as possible. More than one bid was often made to the same company or trust but to different areas of funding that required slightly different criteria. Each of these applications was tailored to suit individual aspects of the project. Some of the companies were earmarked for future applications when these aspects of the project were reaching completion and good visual examples of the nature of the project would be

Terry Jeffrey and friends at a Radio Utopia session. Back row (left to right): Helen Smith, Joy Taylor, Claire Kitto, Jill Patterson. Front row: Vicky Blackwood, Terry Jeffrey, Ann Marie Kitto.
PHOTO: PAUL NUGENT

available. In addition, many environmental funding schemes turned down applications for funding of the project because they saw it as a short-term 'bandage solution' rather than one which was capable of stopping the pollution at source so that it would not happen again.

The newness of the project, its extended time scale and lack of immediate visual evidence of any artistic development within the project slowed down the process of generating additional funding based on precedent. Jozefa Rogocki felt that the problem was that regionally, in terms of other players, Artists' Agency stood alone in its commitment of vision to Seen & Unseen. She believed that if other partners in the project had understood the breadth of vision necessary, it might not have been about a collaborative effort whereby people were just trying to achieve their own piece in the picture.

Jozefa Rogocki acknowledged that she could not directly develop her own personal practice through the project bit that informed the development of future work. Instead, she combined her role as an artist with that of project co-ordinator whenever there was a major event that required the skills of an artist to integrate with the community. This was usually in response to something that had been planned for and was anticipated by the funders as part of a bid made earlier. She took on this dual role either because at the time there was no resident artist in post or because the resident artist chose to have only a limited involvement in any community dimension within a major event. In Jamie McCullough's case the artist chose to remain in the background (as the "woodsman") whereas Helen Smith opted to focus on her own work as part of the event combined with community work that she had initiated (of which Radio Utopia is an example). As a result, Jozefa Rogocki took on the responsibility with Paul Younger and Adam Jarvis for linked schools projects, with Terry and Ian Jeffery for community projects, and with the Hancock Museum and local schools for educational packs to complement the wetland project and related exhibition.

Measuring the success of any project is always a matter of opinion. Key players nearly always believe more could be done and the 'if only' syndrome is a typical reaction. But we need to look at what is actually happening on the ground now if we wish to judge success. The 'local' element of the creation of the wetland is alive and well, and the role of the Trust has been fundamental in determining that success.

Today, the community is managing the day-to-day running of the wetland. A Wetland Newsletter is being produced and a group of volunteers are ensuring that planting takes place, and they are increasingly being asked to show interested groups around the wetland. The listening posts should be in place by the end of 1999 and Craig Knowles, an artist/blacksmith, is designing a metal grille to replace the vandalised glass panel in the wetland jetty. It would appear that the views of some of the funders were unfounded when it came to describing the community as in a worse position than when the project started. The wetland will need constant attention and the members of QHET have shown themselves competent and capable of managing that part of the project.

What members of the QHET cannot do is take Seen & Unseen out into the wider world. We hope that this book will play its part in disseminating information to a wider audience and that the conference in the autumn of 1999 will attract a wide and varied audience.

The truth is that for an innovative interdisciplinary project, Seen & Unseen happened. Maybe it wasn't the perfect way of working together, but all the participants can be proud of setting the standards for others to achieve and better. Collaborative processes have to be the way forward if we are to satisfy the needs of the community and the environment. The goal of achieving a sustainable environment for all depends on co-operation and the sharing of information. Participatory democracy doesn't just happen. It needs vision and the will to *make* it happen.

Seen & Unseen, for all its difficulties has succeeded in providing a new vision for others to emulate.

8 The scientific and engineering context of the Quaking Houses community wetland

Dr. P. L. Younger
Dept of Civil Engineering, University of Newcastle upon Tyne

The concept of passive treatment

Passive treatment of various kinds of waste water has become increasingly widespread in the last two decades of the twentieth century. By 'passive treatment' we mean the improvement of water quality by means of simple, self-sustaining constructed (or natural, but co-opted) ecosystems and associated features (e.g. subsurface reactive drains). It differs from conventional treatment (now increasingly termed 'active treatment') in that passive treatment implies no regular inputs of artificial energy or reactive substances (reagents). The ideal of passive treatment is that we can make a once-for-ever intervention on a site, and then leave the system to take care of itself. In reality, this is seldom completely achievable (for all but the most over-sized of systems), but the pursuit of this ideal at least has the potential to deliver treatment systems requiring minimal maintenance, minimal site security, no power lines or deliveries of chemicals, and no permanent on-site staff.

There are two motivations for passive treatment: one commercial, the other 'ecological'. The commercial motivation is based upon the attractive prospect of making a single capital investment to solve a long-term problem, with little or no requirement for operating expenditure. In the context of abandoned mine sites, for instance, passive treatment might offer a mining company something approximating a 'walk away solution' (notwithstanding the difficulties in truly achieving self-sustaining functionality). The ecological motivation for passive treatment lies in the perceived advantages of adopting a more 'natural' as opposed to 'hard engineering' approach to waste water treatment. Ideally, a passive treatment system should be designed such that it can gradually 'go native', ending up as an integral part of the wider local ecosystem, making a contribution to the maintenance (or even expansion) of biodiversity, as a happy by-product of its primary role of water quality improvement.

The evolution of passive treatment systems reflects this duality of motivations, drawing elements from both conventional, commercial waste water engineering (e.g. Moshiri, 1993; Pescod and Younger, 1999), and from the fields of ecological conservation and restoration (Hammer, 1992; Younger, Large and Jarvis, 1998). These two elements co-exist throughout the existing literature on passive treatment, though it would be fair to say that, on balance, the links to conventional waste water engineering currently remain the strongest. Nevertheless, it should be acknowledged that one of the principal inspirations to the developers of passive treatment has been the observation that many natural wetland systems manage to improve the quality of the waters passing through them (Hedin et al., 1994; Younger, 1997a).

The earliest developments in passive treatment were implemented to improve sewage treatment for small, isolated communities (e.g. Hammer, 1992). Subsequent developments have included applications to agricultural and urban drainage, landfill leachates and mine waters (Moshiri, 1993). Acidic mine waters, such as that found at Quaking Houses, represent something of a special case in passive treatment history, as they differ from most other applications of passive treatment concepts in being concerned with removal of ecotoxic metals and adjusting pH. Most other applications of passive treatment (for sewage, agricultural effluents, urban runoff, landfill leachate and, indeed, net-alkaline mine waters contaminated only with Fe and Mn) are primarily concerned with oxidation of organic matter and/or ferrous iron, and are therefore almost exclusively based on oxidation processes (principally aeration). Only in the passive treatment of acidic mine waters do we see concerted efforts to employ reduction rather than oxidation.

The passive treatment of net-alkaline mine waters contaminated only with iron is achieved simply, by means of oxidation and subsequent precipitation of ferric hydroxides in aerobic reed beds (Hedin et al., 1994). Although certain aspects of this technology remain topics of active research (e.g. Tarutis et al.,

1999), passive treatment of net-alkaline mine waters is now sufficiently well understood that it can usually be implemented with confidence (Hedin *et al.*, 1994; Younger, 1995, 1997b; Laine, 1998). Net-alkaline mine waters contaminated with more soluble metals (particularly Zn and Mn) are more challenging, though progress has also been made recently for these metals in research projects underway at Newcastle (see Nuttall and Younger, 1999).

Passive treatment of net-acidic waters, on the other hand, remains a significant scientific and engineering challenge. This is because of two features common to all 'successful' passive treatment systems for net-acidic mine waters developed to date:

1 They rely on a combination of microbial processes which are not sufficiently well understood that their long-term performance can be predicted with great confidence.

2 They involve subsurface processes of both dissolution (of organic carbon and limestone, most commonly; Hedin *et al.*, 1994; Kepler and McCleary, 1994) and precipitation (of sulphides and hydroxides), which physically alter the geometry (and therefore the hydraulic behaviour) of the passive system over time, in a manner not yet predictable (Hedin, 1997; James *et al*, 1997; Younger, 1997b; Younger *et al.*, 1997).

The net result of these factors is significant uncertainty in the prediction of the long-term performance of passive systems treating net-acidic waters. The uncertainty in the prediction of the long-term performance of passive systems treating net-acidic mine waters appears to be acting as a disincentive to the wider application of an otherwise successful technology. For instance, of five full-scale passive systems constructed by the Coal Authority (CA) since 1996, not one has been for a net-acidic water.

The recognition that passive treatment of net-acidic mine waters is difficult and challenging was one of the key motivations for the Newcastle University scientists to become involved in the Quaking Houses issue. The Quaking Houses wetland represents a virtually ideal 'outdoor laboratory', in which the Newcastle University scientists can begin to probe some of the key questions relating to the application of passive treatment to such 'difficult' waters. The other motivations for the University team were primarily social and environmental (see Younger, 1999), relating to a vision of why and how science (and universities) should be servants of the real needs of the human and natural communities.

The scientific/engineering pedigree of the Quaking Houses wetland

The Quaking Houses wetland, or more particularly the pilot-scale Gavinswelly wetland, was the first system of its kind in Europe. Although Barnsley Metropolitan Borough Council had installed a reed bed to treat acidic leachate at the foot of Dodworth Colliery spoil heap in early 1994 (Bannister, 1997), that system was based on conventional sewage treatment wetlands, and did not incorporate any anaerobic or alkalinity-generating processes (these processes are now being retro-fitted at Dodworth with specialist advice from the Newcastle team). Hence, when the Gavinswelly system was hand-dug and commissioned in the February half-term holiday of 1995, it was truly the first European application of passive, compost-based, sulphate-reduction technology to an acidic mine water.

However, the Quaking Houses wetland did not arise in isolation from the wider world. In particular it drew detailed inspiration, and design rules, from the USA experience, which was (conveniently for us) summarised and codified by the 'three Bobs' (Bob Hedin, Bob Nairn and Bob Kleinmann) of the (now sadly defunct) US Bureau of Mines in Pittsburgh, Pennsylvania (see Hedin et al., 1994). I first became aware of the US experiences with passive treatment in 1993, when a draft copy of the manuscript of the report by Hedin et al. (1994) was mailed to me by Bob Kleinmann. I had previously become aware of the drafting of the report through my involvement as External Technical Review Consultant to the National Rivers Authority (now the Environment Agency) on their Wheal Jane Mine Water project in Cornwall. Already by 1994 there were advanced plans to undertake some passive treatment experiments at Wheal Jane (which were to be designed by Jim Guseck, an employee of the American branch of the lead consultants at Wheal Jane, Knight Piésold and Partners Ltd), and this necessitated my becoming familiar with the US experience. Although the Wheal Jane pilot passive treatment plant was never replaced by a 'full-scale' system, it is testament to the volume of water involved at that site that a system receiving only 2.5% of the total mine water make was substantially larger than the full-scale systems subsequently constructed at other sites in the UK.

With the Gavinswelly wetland functioning, and with the Wheal Jane Pilot Passive Treatment Plant under construction, I visited the US Bureau of Mines team (and particularly Bob Hedin, who had by that time left USBM and established his own business) in Pittsburgh in the summer of 1995. This afforded me the chance to see at first hand some of the best documented and longest established passive systems in the world, the best example being the Howe Bridge system in Clarion County, PA. It also gave me the chance to learn how the so-called SAPS (Successive Alkalinity Producing Systems; Kepler and McCleary, 1994) were beginning to replace simple compost wetlands as the systems of choice for acidic waters like that at Quaking Houses.

A SAPS comprises a bed of limestone gravel overlain by a compost layer. Water is forced to flow downwards through the compost and the limestone in turn. In the compost layer, the water is stripped of dissolved oxygen and its ferric iron is converted into the ferrous form (which renders it incapable of clogging the underlying limestone bed with ochre). Dissolution of the limestone raises the pH and puts alkalinity (primarily bicarbonate) in the water. Subsequent oxidation of the water after leaving the limestone bed results in rapid precipitation of ochre, with the pH being maintained at an elevated value thanks to the buffering afforded by the alkalinity. SAPS systems require a higher level of engineering than simple compost wetlands, and also need a reasonable drop in head (i.e. net water level) across a prospective site. We always knew that head on the Quakies site would be tight, but we considered that excavation of a deep pond and back-filling with limestone and compost might give us scope for Europe's first SAPS. We persisted in this hope until 1997 (Younger *et al*, 1997) when the site investigations revealed that the site was underlain by unrecorded colliery washery slimes, which were so acid-generating that they precluded wholesale excavation. (In the end, the Newcastle team *did* design the first SAPS in Europe, but this was constructed at the Pelenna III site, Tonmawr, South Wales; see Ranson and Edwards, 1997; Younger, 1998b).

It was in the detail of the response to this unforeseeable design constraint that the Quakies wetland was turned from a foiled attempt at a SAPS into a system with further novel attributes not previously implemented elsewhere. Of course, the main unique characteristic of the Quaking Houses wetland is the community-artist-scientist linkup, so amply documented in the foregoing chapters. Yet even as a technical entity the wetland has a number of unique attributes. The first was the bund itself, which was formed by painstakingly compacting grey, powdery pulverised fuel ash (PFA) to form the outer barrier, central dividing bar and islands of the wetland. The suggestion to use PFA was made by my senior colleague, Iain Moffat, a specialist dam engineer and Senior Lecturer in the Department of Civil Engineering at the University of Newcastle. Iain also gave Adam Jarvis invaluable advice on the critical width-to-height ratio of the bund, advice which has since been borne out in faultless performance (with neither structural failures nor leaks) to date. The use of PFA allowed (or rather, required!) the participation of volunteers from Quaking Houses village, operating the compacting machine for many hours in the summer heat. Although PFA has found a number of practical uses, the most well-known being in the manufacture of breeze blocks, this is as far as we can ascertain the first time it has been used in constructing a mine water treatment wetland. The alkaline properties of PFA made it particularly suitable for this application, as it offered the prospect of a 'pH holiday' for the wetland until such time as the microbial communities required for the principal sulphate-reduction process could become fully established.

Another novelty of the Quakies system is the blend of compost substrates used. Apart from horse manure and straw (as used in the pilot system), we used cow manure and straw (from the University's teaching and research farms in Northumberland) and composted municipal waste from Castle Morpeth Borough Council's waste composting facility (a suggestion made by Professor Ken Anderson at the University). As far as we can tell from our sampling to date, all three composts are performing very well.

In essence, the Quakies system is an example of how one source of waste (in this case, solid wastes such as composted municipal waste and PFA) can be used to 'cancel out' another (acidic mine water). Parallel developments in South Africa, where tannery wastes and sewage have been co-treated with acidic mine waters (Rose et al., 1998), demonstrate the wider applicability of the kind of lateral thinking which characterises the engineering design at Quakies.

Through a combination of unplanned delays in construction, and the planned timing of the first national conference on 'Minewater Treatment Using Wetlands' (held at Newcastle University in September 1997 under the auspices of the Chartered Institution of Water and Environmental Management; see Younger, 1997b), it happened that Bob Hedin was able to visit the Quakies system under construction, and give his unofficial benediction on the traditional and innovative aspects of the new wetland.

Wetland treatment processes and performance in scientific terms

The foregoing chapters have described well the theory and practice of mine water treatment at Quakies, which may thus be summarised very briefly here as (see Younger et al, 1997, and Jarvis and Younger, 1999):

1 Removal of acidity, raising of pH and generation of alkalinity by means of

limestone dissolution (hundreds of kilograms of limestone cobbles are mixed with the substrates, and stacked in a thick bed at the end of the second pond).

2 Removal of acidity, raising of pH and generation of alkalinity by means of bacterial sulphate reduction within the compost substrate.

3 Trapping of iron as a sulphide within the substrate, by reaction of the iron with sulphide produced by the sulphate-reducing bacteria.

4 Trapping of aluminium as a hydroxide ($AlOH_3$) when the rise in pH makes this phase insoluble.

The synthesis of previous US experience put together by Bob Hedin (*et al*, 1994) had led us to expect average rates of acidity removal amounting to about 7 g.d^{-1}.m^{-2}. In the event, the pilot (Gavinswelly) wetland achieved an average of 9.6 g.d^{-1}.m^{-2} (Younger *et al.*,1997). We were not so rash as to expect this happily elevated rate to be matched in the full-scale system, and yet we were proven too cynical, as the full-scale system achieved an average rate of 10.4 g.d^{-1}.m^{-2} (Jarvis and Younger, 1999). As the graphs in the Appendix show, the performance of the system fluctuates over time. It is a matter of simple observation that the wetland seldom looks the same on two consecutive visits. Sometimes a sheen of aluminium hydroxides will cover most of the surface. At other times, a 'slick' of floating bacterial colonies (*Leptothrix spp.*) will lie on the water surface. Often, the water will be clear, and a greenish-tinged substrate will be clearly visible, pock-marked where streams of gas bubbles (the sure signs of microbial activity down below) leave the substrate via mini underwater vents. Such variation is only to be expected in a semi-natural (and presumably still acclimatising) ecosystem. Certainly the quality of the incoming water varies considerably over time, and the treatment process appears to be at least first-order with respect to influent concentrations. Variations in temperature and nutrient availability are inevitably reflected in changes in bacterial metabolism, and hence in the rate of sulphate reduction. At the time of writing, the details of these processes, and their ramifications for the long-term performance of the wetland, are key foci of research (proposed and in progress) at Newcastle University.

'Little Quakies is turned into a giant world!'

At the time of the Wounded Knee stand-off in 1973, the Lakota people were amazed and encouraged by the speed and vigour with which their protest against the injustices of US rule was taken up by sympathisers worldwide.

What had started as a local protest against the excesses of sadistic policemen quickly transformed itself into a beacon of hope for oppressed peoples the world over. Wallace Black Elk expressed the pleasant surprise felt by the Lakota people thus: 'Little Wounded Knee is turned into a giant world' (see *Akwesasne Notes*, 1974, p 91). Whilst not wishing to compare our humble achievements at Quakies with the re-awakening of Indian radicalism in the Americas which Wounded Knee'73 engendered, I can't help recalling Wallace Black Elk's exclamation at times. For Wounded Knee is a quiet little corner inhabited by people the world thought it could ignore; not so very different from Quakies, after all. When I contemplate the international interest that Quakies has aroused in the mining and environmental management fields, it is sometimes also a source of wonder to me that our 'little bit wetland' is emblematic to so many people around the world.

First and foremost, of course, was the winning of the 'Conservation Engineering' and 'Overall Winner' categories of the UK Conservation Awards 1998, as described in the earlier chapters. This prize brought us not only national publicity and much-needed cash to support the early monitoring of the full-scale wetland, but also took Alan McCrea and myself to Istanbul, where Quakies represented the UK in the European Conservation Award finals. Less obvious, but no less important in its way, has been the steady stream of international scientific visitors to Quakies, from the days of Gavinswelly onwards. Indeed, Almudena Ordoñez Alonso (now a lecturer at the School of Mines in Oviedo, Spain) worked day after day with Adam Jarvis and the Quakies volunteers on the construction of the full-scale wetland. Subsequently, in one memorable week in late 1997 shortly after the full-scale wetland was first commissioned, we had two separate delegations from Spain, both of whom returned home (one group to Galicia in the north-west, the other to Toledo in the south) to build mine water treatment wetlands of their own. Most recently, partly in recognition of the experience gained at Quakies, I was invited to advise the Junta de Andalucía on the possible role of wetlands in the long-term remediation of Europe's worst ever mine spill, at Aznallcóllar, near Seville. Quakies wetland has now been described in publications in the Spanish language (Ordoñez *et al*, 1998; Younger, 1998a), and in conference proceedings published in Johannesburg, South Africa (Younger,1998b). Research collaboration between the Newcastle University team and the team from Rhodes University, South Africa, who developed the systems for co-treatment of mine waters and sewage (Rose *et al*, 1998) is now burgeoning, with scientists from Cape Province scheduled to spend several months each year working on the microbiology of the Quakies system over the first few years of the twenty-first century.

Little Quakies is not doing so bad at turning into a giant world!

References

Akwesasne Notes, 1974, *Voices from Wounded Knee, 1973, in the words of the participants*. Published by Akwesasne Notes, Mohawk Nation, via Rooseveltown, New York 13683. (ISBN 0-914838-01-6). 263pp.

Bannister, A.F., 1997, 'Lagoon and reed-bed treatment of colliery shale tip water at Dodworth, South Yorkshire'. In Younger, P.L., (editor), *Minewater Treatment Using Wetlands. Proceedings of a National Conference held 5th September 1997, at the University of Newcastle, UK*. Chartered Institution of Water and Environmental Management, London. pp 105–122.

Hammer, D.A., 1992, *Creating Freshwater Wetlands*. Lewis Publishers, Boca Raton, Fla. 298pp.

Hedin, R.S., 1997, 'Passive mine water treatment in the eastern United States' in Younger, P.L., (editor), *Minewater Treatment Using Wetlands. Proceedings of a National Conference held 5th September 1997, at the University of Newcastle, UK*. Chartered Institution of Water and Environmental Management, London. pp 1–15.

Hedin, R.S., Nairn, R.W. and Kleinmann, R.L.P., 1994, *Passive Treatment of Coal Mine Drainage. US Bureau of Mines Information Circular 9389*. US Department of the Interior, Bureau of Mines, Pittsburgh, PA. 35pp.

James, A., Elliott, D.J., and Younger, P.L., 1997, 'Computer aided design of passive treatment systems for minewaters.' In: Younger, P.L., (editor), *Minewater Treatment Using Wetlands. Proceedings of a National Conference held 5th September 1997, at the University of Newcastle, UK*. Chartered Institution of Water and Environmental Management, London. pp 57–64.

Jarvis, A.P., and Younger, P.L., 1999, 'Design, construction and performance of a full-scale wetland for mine spoil drainage treatment, Quaking Houses, UK.' In *Journal of the Chartered Institution of Water and Environmental Management* (in press).

Kepler, D.A., and McCleary, E.C. Successive Alkalinity Producing Systems (SAPS) for the Treatment of Acidic Mine Drainage. *Proceedings of the International Land Reclamation and Mine Drainage Conference and the Third International Conference on the Abatement of Acidic Drainage*. (Pittsburgh, PA; April 1994). *Volume 1: Mine Drainage*. pp 195–204. 1994.

Laine, D.M., 1998, 'The treatment of pumped and gravity minewater discharges in the UK and an acidic tip seepage in Spain.' In *Proceedings of the International Mine Water Association Symposium on 'Mine Water and Environmental Impacts'*, Johannesburg, South Africa, 7th–13th September 1998. Volume II: pp. 471–490.

Moshiri, G.A., 1993, *Constructed Wetlands for Water Quality Improvement*. Lewis Publishers, Boca Raton, Fla. 632pp.

Nuttall, C.A., and Younger, P.L., 1999, 'Zinc removal from hard circum-neutral mine waters using a novel closed-bed limestone reactor.' *Water Research* (in press).

Ordoñez, A., Younger, P.L., and Jarvis, A.P., 1998, 'Depuración de agua de mina mediante humedal en UK.' In *Proceedings of the Xth Congreso Internacional de Minería y Metalurgia* (International

Congress of Mining and Metallurgy), Valencia 1st–5th June 1998. pp. 307–317.

Pescod, M.B., and Younger, P.L., 1999, 'Sustainable Water Resources.' (Chapter 3). In Nath, B., Hens, L., Compton, P., and Devuyst, D., (editors), *Environmental Management in Practice: Volume 2. Compartments, Stressors, Sectors.* Routledge, London. pp 55–73.

Ranson, C.M., and Edwards, P.J., 1997, 'The Ynysarwed experience: active intervention, passive treatment and wider aspects.' In Younger, P.L., (editor), *Minewater Treatment Using Wetlands. Proceedings of a National Conference held 5th September 1997, at the University of Newcastle, UK.* Chartered Institution of Water and Environmental Management, London. pp 151–164.

Rose, P.D., Boshoff, G.A., van Hille, R.P., Wallace, L.C.M., Dunn, K.M., and Duncan, J.R., 1998, 'An integrated algal sulphate reducing high rate ponding process for the treatment of acid mine drainage wastewaters.' *Biodegradation*, 9, pp 247–257.

Tarutis, W.J., Stark, L.R., and Williams, F.M., 1999, 'Sizing and performance estimation of coal mine drainage wetlands.' *Ecological Engineering*, 12, pp 353–372.

Younger, P.L., 1995, 'Hydrogeochemistry of minewaters flowing from abandoned coal workings in the Durham coalfield.' *Quarterly Journal of Engineering Geology*, 28, (4), pp S101–S113.

Younger, P.L., 1997a, 'The future of passive minewater treatment in the UK: A view from the Wear Catchment.' In: Younger, P.L., (editor), *Minewater Treatment Using Wetlands. Proceedings of a National Conference held 5th September 1997, at the University of Newcastle, UK.* Chartered Institution of Water and Environmental Management, London. pp 65–81.

Younger, P.L., (editor), 1997b, *Minewater Treatment Using Wetlands. Proceedings of a National Conference held 5th September 1997, at the University of Newcastle, UK.* Chartered Institution of Water and Environmental Management, London. 189pp.

Younger, P.L., 1998a, 'Tratamiento Pasivo de Aguas de Minas Abandonadas en Gran Bretaña.' In *Procedimientos de la Reunión Científico-Técnica Sobre el Agua en el Cierre de Minas.* Dpto de Explotación y Prospección de Minas, Escuela Superior de Ingenieros de Minas, Universidad de Oviedo, Asturias, España. 16pp.

Younger, P.L., 1998b, 'Design, construction and initial operation of full-scale compost-based passive systems for treatment of coal mine drainage and spoil leachate in the UK.' In *Proceedings of the International Mine Water Association Symposium on 'Mine Water and Environmental Impacts'*, Johannesburg, South Africa, 7th–13th September 1998. (Volume II). pp. 413–424.

Younger, P.L., 1999, 'Restless waters of the Durham Coalfield: Pollution risks and popular resistance.' *Bands and Banners*, 1, pp 19–21.

Younger, P.L., Curtis, T.P., Jarvis, A.P., and Pennell, R., 1997, 'Effective passive treatment of aluminium-rich, acidic colliery spoil drainage using a compost wetland at Quaking Houses, County Durham.' *Journal of the Chartered Institution of Water and Environmental Management*, 11, 200–208.

Younger, P.L., Large, A.R.G., and Jarvis, A.P., 1998, 'Considerations in the creation of floodplain wetlands to passively treat polluted minewaters.' In Wheater, H. & Kirby, C. (eds.), *Hydrology in a Changing Environment. Proceedings of the International Symposium organised by the British Hydrological Society, 6th–10th July 1998, Exeter, UK.* Volume I. pp 495–515.

9 Art in Seen & Unseen: Context and evaluation

Malcolm Miles
Oxford Brookes University

The previous chapters have given a history of the wetland and its relation to the community of Quaking Houses, and evaluated it scientifically. This chapter has two further aims: to place Seen & Unseen in a context of environmental art, recently and further back in history; and to ask what was achieved by the involvement of artists in the project.

There is a long history of art about the land. Traditionally, artists have found inspiration in natural beauty, expressed in landscape painting and other forms. Recently, some artists have worked in remote places making art from the materials of the site itself; others have collaborated with other professionals and communities to explore ecological issues and futures, or intervened in practical ways to reclaim land from industrial pollution.

So, what do artists contribute to initiatives such as the creation of a wetland? Does their intervention lead them beyond making the sorts of things most people think of as art? And should collaborative work be evaluated according to aesthetic or social criteria? As artists become increasingly disillusioned with the art market, turning instead to social settings for their work, such questions affect art's future direction. But they also affect the development of environmental projects. Money from arts funding bodies may cover the costs of introducing art into a project, and art tends to gain media attention. But does it contribute to the realisation of the aims a project sets out to achieve? Does art play an integral part, enhancing the quality of a project as a whole, lending the experience a deeper or more enduring meaning? Or does it remain dependent on questions of style, taste and market value which only experts can decide? If art does contribute to a better world, does it retain its quality as art? These are complex questions, and the chapter sets out a context in which to consider them before confronting them.

In the end, art, unlike science, does not deal in right answers. It has been described as a pathless land. Whilst there is a loose consensus among arts professionals as to what counts as contemporary art, and many individuals express ideas as to what they like (or, more often, don't like), it is hard to say exactly how art is successful. Sometimes the most imaginative solutions emerge when conventional approaches are questioned and reconstructed through working with people from different backgrounds or professions. Part of the interest of Seen & Unseen is that art was involved throughout, and the project provides a vehicle for further awareness of these issues.

Environmental art—a history

Art has many histories. Developments in art have more than one possible starting point. For environmental art, the 1690s is as good a time to begin as the 1960s, although most environmental artists today would relate their work to recent precedents. Yet environmental art inherits a context of art about the land in cultures in which the landscape has a special place.

When Dutch artists of the seventeenth and eighteenth centuries took the landscape of their country as a subject for art, setting aside the convention that good art required a historical or mythological story, they demonstrated a concern for the land. Much of their recently independent state was reclaimed from the sea (another favourite subject for a maritime, trading nation), and its depiction was part of the making of a national culture. When Constable painted scenes of rural Suffolk, or sketched the weather on Hampstead Heath, he celebrated Nature. To him it was bountiful in the golden corn, dangerous in the unpredictability of the storm. Constable was concerned to show a relation between people and the land, and to draw attention to the poverty of rural labourers. His work, such as The Haywain, and Millet's Angelus (engravings of which were found in many Victorian middle-class houses), have become part of a kind of English culture, but perhaps not in ways either artist intended. Still, landscape painting, and the poetry of Wordsworth, has contributed to a feeling that the distinctiveness of its landscape is part of what a nation values. The industrial agriculture which uproots hedgerows is produced by a society whose members also flock to exhibitions of landscape art. A challenge for art today is to point out the contradiction.

An allusion to the destruction of the natural environment is found in nine-teenth-century North American painting, as artists followed the railroads into the west. Artists such as Frederick Church and Thomas Cole charted the vastness and the intricacies of the American scene, but Asher Durand's *Progress* (1850) shows the triumph of industry and transportation over the land. Durand linked the wilderness with native Americans, seeing both as defunct in the face of a superior civilisation. Meanwhile, in imperial Britain, Darwin's theory that a species survives according to how well it adapts to its environment was twisted into the notion of survival of the fittest. Today, Durand's painting seems less heroic, though the pioneer myth survives in the North American attachment to log cabins. Artists have been described, too, since the Romantic movement, as lonely frontiersmen exploring the primor-dial forests of human emotion.

The above examples from Holland, England and North America, in their respective traditions, take the land and people's relation to it as subject matter, framing it as land-scape. This is not new. But whilst earlier artists described landscape, environmental artists in the 1960s and '70s re-constructed it phys-ically in the form known as earthworks (or land art), often working in remote locations. American artists such as Nancy Holt, Michael Heizer and Robert Smithson found a new expressive potential in the relation of land, sky and constellations, and the human spectator. In Britain, Andy Goldsworthy has made art from twigs, leaves, water, ice, stones, and dust, and Richard Long has developed a practice based in walking the land. A second difference from historical art is that the land is seen today more as a site of clashing values, of issues which are more social and political than aesthetic. Perhaps it always was like that, as when Dutch artists defined a national culture through a national terrain. But today's eco-artists go beyond the depiction of the land, even the aims of land art, being concerned more to reclaim than reconstruct landscape. In this way they respond to global and local, rather than national, questions. In the face of corporate greed and the power games of states, artists seek through collaborations with environmentalists and local people to find ways to sustain rather than destroy life on Earth. This kind of art does not produce the paintings and sculptures which, after the death of the artist, sell at auction for millions of pounds or dollars. Whilst some ecologically-concerned artists use traditional skills, most work conceptually. This is part of a wider tendency since the 1960s called the de-materialisation of art. The new art is process-rather than object-based, the artist a researcher.

Why did this happen? Artists, since the late 1960s, have increasingly refused to make objects to serve an art market the values of which are alien to them. For many it seems obscene that a painting of sunflowers by an artist who died in abject poverty should be sold for a sum adequate to build a hospital. This was also the era of protest against the American war in Vietnam, civil rights marches, and women's liberation. Around this time, some artists saw an alternative to the art market through work as facilitators for art in community settings. Community-based art gave a voice to minority groups marginalised by the affluent society. The visual quality of murals, banners and lantern parades was not irrelevant, but success was seen in a quality of social engagement and the extent to which people who were not artists could find creativity and identity through art, rather than measured by the taste of an elite. The history of community involvement, like that of land art, is the legacy of today's environmental artists. Mierle Ukeles, for example, became un-funded artist in residence with the garbage collectors of New York in 1970, and has continued to investigate attitudes to the production, handling and recycling of waste. If there is a third inheritance, it is the work of artist Joseph Beuys, whose dialogues sought to liberate and who played a role in the founding of the Green Party in Germany.

Environmental art today has four main aspects: art which reclaims polluted or damaged land; art which draws attention to threats to habitat and diversity of species; art which works practically to heal the land of pollution; and art which foresees alternative futures. We now look at each of these categories.

The varieties of environmental art today

One of the earliest cases of land art, often overlooked, is Harvey Fite's *Opus 40*, a sculpting of a disused bluestone quarry in the Catskills, near Woodstock, in New York State. Fite's intention was to take out stone for his sculptures, then use the site to display them. But for forty years until his death in 1976 the quarry became an artwork in itself, a blot on the landscape transformed into an aesthetic space. But not all cases of reclamation art have been problem-free. Robert Morris, working with an abandoned gravel pit near Seattle in 1979-80, contributed to the erosion of the site by removing trees to emphasise its industrialisation. Questions have also been asked about Robert Smithson's work, such as *Spiral Jetty* of 1970 in Utah, and his unrealised proposals for mining wastes in Colorado. British artist Hamish Fulton argues that art which re-constructs the land in this way lacks respect for it. Were these artists captivated by the power of modern earth-moving equipment? Or was their work an effort to make a post-industrial landscape?

Other artists took a more gentle approach. Alan Sonfist recreated the original landscape of a vacant lot in south Manhattan, planting native trees. Doug Hollis made a series of wind-activated sound sculptures at the National Oceanic and Atmospheric Administration site in Seattle in 1983. And Richard Long made perhaps the least violent of interventions by walking the land, at most re-placing a few stones, and documenting the process in photography and text. Long's sites ranged from Bolivia to the Himalayas, as well as rural Britain. He says of his work that stones are the material of the Earth, and he hopes the Earth has a future. Fulton developed a similar strategy, walking in California, India and elsewhere. Through photographs taken during his walks, Fulton protests against urban alienation from the natural world. Alienation is the subject, too, of American artist Dominique Mazeaud's work. Once a month, from 1987, she walked the dry bed of the Rio Grande in Santa Fe, removing litter. Mazeaud wrote a diary of the walks, reflecting on the sacredness of water and, as all rivers lead to seas, and seas provide the moisture which becomes rain, its aptness as a metaphor for the connectedness of all life.

Responding to the extinction of species due to the loss of natural habitats, the environmental charity Common Ground has worked with artists in rural places since 1985. The projects were begun on the initiatives of local landowners and communities, but co-ordinated nationally. Amongst them was a commission for stone carvings in Dorset by Peter Randall-Page. Three spiral shell forms are set in dry-stone niches on a bridleway near Lulworth. The stone used for carving is a local limestone composed of fossilised shells. Next to fields where barley has grown for two thousand years, the sculptures mediate between the scales of landscape, people and small creatures in the grass [see Plate 10.1]. In *Still Life*, for European Year of the Environment in 1987, Randall-Page carved three monumental life-forms from the list of endangered species, setting lines by Victorian nature poet and Jesuit Gerard Manley Hopkins on the plaque:

> "What would the world be, once bereft
> Of wet and wilderness? Let them be left,
> O let them be left, wilderness and wet:
> Long live the weeds and the wilderness yet."

A small number of art projects have piloted practical ways to heal land polluted by industrial waste, or find sustainable patterns of development. The London-based group PLATFORM has worked since 1989 to raise interest in, amongst other things, London's buried and neglected rivers, and is generating renewable energy from one of them to light a school. PLATFORM presented their approach at the brainstorming day for Seen & Unseen. Another example is Mel Chin's *Revival Field*, begun in 1989. Chin used plants which take up toxins such as cadmium from the soil to create a circular space of reclamation on a toxic waste site near Minneapolis. Chin worked with an agronomist, finding corn, bladder campion and pennythrift the most useful plants. A second site was established in Palmerston, Pennsylvania in 1992, and two more in the Netherlands. Chin's projects have proved the effectiveness of this green technology. Only in the art world has he encountered the rebuke that his work ceases to be art. His response is that he uses the traditional method of carving—removing material to reveal form—though the material removed is industrial waste and the form revealed a revitalised Earth. A project which unites the healing aspect of Chin's work with a concern to extend bio-diversity and reclaim derelict land as public space is *Nine Mile Run* in Pittsburgh. Tim Collins, Reiko Goto and Bob Bingham, of the Studio for Creative Inquiry at Carnegie Mellon University, are collaborating with city authorities, geologists, water engineers, the Rocky Mountain Institute and developers to reclaim a 230-acre site used from 1922 to 1970 as a slag dump as a post-industrial landscape [see Plate 10.2]. A stream runs through the valley, identified by Frederick Law Olmsted Jr (son of the planner of Central Park) in 1910 as a site of wooded slopes and special beauty, traces of which remain. The Studio has set up a trailer near the stream bed which runs through the site as a base for community links, and acts as a bridge between community representatives, developers and city authorities.

The most conceptual area of ecological art, pioneered by Californian artists Helen Mayer Harrison and Newton Harrison, gives form to visions of alternative futures. In 1998, the Harrisons worked with students at Manchester Metropolitan University to map an ecological domain between the Humber and the Mersey. They held an open studio in which to dialogue with planners, architects, economists, conservationists and ecologists. Plotting the likely outcomes of free market development as black cancer-like growths on the map, they contrasted this with more optimistic futures including zones of protected bio-diversity. They found that the organic waste of the domain's nine million human inhabitants could be redistributed over the 900 square miles of farmland to meet most of the need for soil enrichment, allowing chemical fertilisers to be almost set aside.

Where does Seen & Unseen fit in these histories? With PLATFORM's work it shares a concern to link ecology with public engagement; and with Mel Chin's it shares the development of practical methods to heal a ravaged Earth. The wetland also has much in common with *Nine Mile Run*. Both projects, developed independently of each other, involve practical steps to reclaim natural resources from industrial pollution. Both involve collaborations between people from a range of cultural and scientific professions and local communities. And both aim to produce a new public space. The Pittsburgh project is bigger, though the small scale of the wetland allows a more intimate relation to the community of Quaking Houses. So, Seen & Unseen, a unique project in many ways, is also part of a new and international history of work which integrates art, science and the energy of communities in building sustainable futures.

Art in the wetland

Three artists have been involved in the wetland at Quaking Houses. Jamie McCullough worked with the scientists and engineers to resolve some of the questions arising in the project's early days. Helen Smith was artist in residence whilst the wetland took shape. She contributed to the design of the site, its gradients and water flows, and the walkway giving access to it, as well as working with new technologies of information and communication to create dialogues between local people. Her proposal for a set of *Listening Posts* is ground-breaking, both as art and as a way to integrate local voices with a popular understanding of science. An exhibition of work derived from her experiences of the wetland was held at the Hancock Museum in Newcastle in 1998. Finally, introducing a dimension of craft, basket weaver Lee Dalby made small-scale sculptures in willow and engaged local young people in basket-making workshops. More recently, he has made works in living willow which will be maintained, following further workshops, by local people.

Community arts projects have often suffered from a flawed understanding of what constitutes a community, or a lack of common concern. McCullough recognised that at Quaking Houses there was a rare case of real community initiative. He also addressed the problem that professionals in different fields sometimes speak what seem different languages, seldom accessible to ordinary people. Discussions in the planning stage were wide-ranging and time-consuming, and perspectives did not always coincide. But McCullough injected "a little wonder and imagination", as he put it, into the early stages of the wetland.

The next phase was the design and construction of the wetland. The chemistry of water purification could work in square concrete tanks, but the organic processes suggest landscape. Helen Smith, as artist in residence, worked as part of the project team to design the wetland's undulating curves, its islands and areas of planting to encourage wildlife. Whilst the wetland is an artificial landscape (like the lakes of nineteenth-century landscape parks), it sits well in the site. When the vegetation has matured, it will seem as if it has always been there. Smith ensured that as many members of the community as possible could enjoy the wetland by designing a walkway wide enough to support a wheelchair. She did not restrict her work to designing on paper, but took part in the physical work of sinking wooden supports and digging out the site [see photo, p85].

The wetland is a community project, and central to its success is community ownership, which is more than physical access. Smith facilitated a kind of metaphorical access, working with young people to train them in information and sound technologies, utilised in two hour-long broadcasts on Sunderland University's Radio Utopia [see photo, p106]. The broadcasts gave people an opportunity to share feelings about the locality, offered different generations a way to work together, and integrated data on the project with local narratives. This formed the basis for her proposal for a series of *Listening Posts*. Made at the University of Northumbria, they combine information technology with imaginative soundscape, and transmit a mix of local narratives. Facts on water purification monitored by local volunteers are interspersed with material derived from Radio Utopia and more recent recordings with local people of all ages.

The *Listening Posts* do not look like the kind of art found in museums. They are part of a tendency in contemporary art towards research, and the most innovative outcome of Smith's involvement. But then, Leonardo da Vinci filled his sketchbooks with observations of how water flows and designed flying machines. The idea of artist as researcher is not new, nor that of artist as communicator. Jean-Jacques Rousseau proposed in the eighteenth century that art and literature are ways to spread new ideas. Smith has brought the role of artist as communicator into an age of democracy and information technology. She states:

"A framework was provided for my work [by] the need for access by the environmental campaigners to the processes of the scientists. What I have tried to do is play with that on behalf of and in collaboration with the village, as well as working with the scientists on tying my contribution into the needs and functions of the wetland."

These relationships were articulated in an adapted Victorian Davenport desk, in which interactive audio-visual equipment is installed and each drawer offers an aspect of the exploration the project has involved—exhibited at the Hancock Museum in 1998 [see Plate 14].

Smith's decision to focus on interactive communication was innovative. But there is, too, a place for work in more traditional media. Lee Dalby created a fossil-like form in willow to place in the mud of the wetland, reminiscent of the coalfield (a fossil fuel) under the site [see Plate 11]. Fossil-forms suggest geological time. Millions of years ago, the coalfield was a forest. The trees then became carboniferous sediments under a sea, eventually compressed into coal under layers of earth, which led to mining and the pollution which the wetland now cleanses. Dalby's sculptures of living willows will sprout and grow over a period of years, adding to the integration of a made landscape in the natural land. His basket-making workshops and training of local people in cutting and maintaining the newly planted willow works add to community ownership of the project.

Evaluation—art, science and locality

Each artist made a particular contribution, but how can the role of art in the wetland as a whole be evaluated? This final section offers some reflections on how the involvement of artists might be seen. It argues that art was integral to the project, and that its contribution has radical implications for wider patterns of thought and attitudes to the land. Finally, it summarises the benefits of the artists' work, returning to the questions posed at the beginning of the chapter.

"Revolutionary project helps clean stream" proclaimed a local newspaper—"revolution" sounds world-shaking, yet "stream" conjures an image of an intimate landscape. The conjunction sums up the project's local response to the global problem of pollution. But why is Seen & Unseen revolutionary? The answer has three aspects: the integration of knowledges from art, science and the local community; the integration of functional effectiveness, access and beauty; and the innovative use of technologies of communication. In each of these the role of the artists has been central.

To take the first aspect: art, science and locality represent three kinds of knowledge with different methods and philosophies. Science in the modern world emphasises problem-solving and seeks the kinds of certainties found in geometry and mathematics. Seeing all forms of matter as of equal interest, it experiments to see how they behave in different conditions and finds laws, like gravity, which operate under all known conditions. Just as two plus two equals four, so apples always fall downwards off the tree—Newton sorted that out. Science is sometimes called value-free because it seeks new knowledge without moral or religious censorship. The scientific revolution of the seventeenth and eighteenth centuries established this, though today genetic engineering throws the desirability of that freedom into doubt and the genetic modification of seeds demonstrates a fusion of science and free-market economics. Art, on the other hand, tends to ask more questions than it answers; it poses problems, including the problem of what art is. Contemporary art deals as much with knowledge as skills, and the skills it develops are in the construction of meaning. Like pure science, art requires no use beyond being art. Art since the Romantic period, again like science, is free of moral and religious interference, and artists are seen as free from conformity to social norms. But art is not value-free because it either makes claims to express human values, or states the value of an autonomous aesthetic dimension.

These differences do not mean there are, as writer C. P. Snow thought, two separate cultures of art and science. On the contrary, research in particle physics and cosmology involves imagination, whilst projects for art linked to medicine, and research in visual perception, bring art and science together. But both are given a new focus by adding the third kind of knowledge: that of a community.

The word community is over-used, and in the political rhetoric of "the international community" has no meaning. Where community does mean something, it denotes a group of people with something more than a label in common. In urban and regional planning, dwellers in a given neighbourhood are increasingly brought into debates. Just as experts, including water engineers and artists, have knowledge on water engineering and art, so local people are experts on living where they do. The Quaking Houses

Environmental Trust was instrumental in the creation of the wetland, and although the success of organic methods of cleansing polluted water depends on scientific knowledge, the involvement of local people was essential to the sense of emotional ownership evident in the project. Without that it is doubtful whether it could have happened at all, money have been raised, or regular monitoring and site maintenance carried out. The artists, particularly Helen Smith, made a major contribution to integrating these three kinds of knowledge and thus to the project as a whole.

To take the second aspect: the separation of the functional from the aesthetic goes back to classical times, when the arts and philosophy assumed a privileged position compared to manual skills like building ships or baking bread. In earlier societies, those activities had sacred connotations, but by the time of the Parthenon, the division of society into the free and the unfree was reflected in a hierarchy of knowledges in which the useful was the lower and the useless the higher. From those roots comes our society's separation of art from everyday life. Since the eighteenth century, art has taken on an added role as purveyor of dreams in an age of religious decline, and in this respect has extended to all classes. What was once a heavenly reward after a life of piety and suffering is now found in the gallery, the romantic novel or the cinema. Art as escape, and the illusions of consumer culture, discourage revolt. To refuse the separation of art from ordinary life, then, is a revolutionary act in the literal as well as metaphorical sense. A wetland might not seem exactly a danger to the state, but in as much as it is the product of a collaboration between artists, scientists and the community, making a place which fuses beauty with use, it redefines given values and shows what even a small community can do to change the world. The focus on public access is equally important. The more people are attracted to visit the site, the more the scientific model it demonstrates becomes widely known. But also, the creation of a site of beauty brings that dimension, in a practical and directly experienced way, into the everyday lives of local people.

The third revolutionary aspect of the wetland is the extension of the artist's role as communicator. Over the past decade, new technologies of processing information have produced touch-screen cash-points, computers in schools and shopping via the internet. The technologies are becoming cheaper and easier to use, and cyber-space is almost as accessible, in affluent countries, as the street. But nearly all the advances have been driven by the needs of industry and commerce, such as the need for 24-hour dealing in

Jim Buchannan, Willow Ball, at the Earth Centre, South Yorkshire.
PHOTO: MALCOLM MILES

shares and currencies, or have military origins. The *Listening Posts* translate
new technology into an everyday application. Potentially dry scientific data
becomes part of a dialogue which embodies the relation of the wetland to the
community whose efforts brought it about. The involvement of local people
in the project can be passed on to visitors and younger generations. This is
part of sustainability, as much as environmental science. As someone said:
"You can't just design sustainability, you have to make it happen"; and that
entails the glue of social ownership.

These three, overlapping aspects of the project demonstrate how central
art has been to its success. Just as landscape paintings and poetry help people
find their own meanings in the land (in a way the tourist poster or the post-
card view usually do not), so elements such as Radio Utopia and the *Listening
Posts*, in today's technology, create that engagement. In taking on a social and
environmental role, the artists have not relinquished their imaginations or
practical skills. But they have been content to let their work be evaluated

within a wider purpose of building a better world. The wetland is less spectacular (and expensive) than Antony Gormley's *Angel of the North*; but the criteria through which to evaluate it are not its conformity to prevailing fashions, nor its monumentality. More important are the qualities of social relatedness and ecological healing.

This is not to abandon beauty—visions of a future free of pollution can be as beautiful as the land itself. And the wetland is beautiful, in the way of landscape which sinks into people's minds [see Plate 16]. The evidence so far is that many local people visit the wetland. For some this land is itself infused with memories. For others, particularly the young, the memories are of their involvement in an aspect of the project, such as the basket-weaving workshops or Radio Utopia. In a society in which crime and vandalism are widespread, and the ethos of corporations (and sometimes the state) no less destructive, it might have been expected that a project as fragile and open as the wetland, a fifteen-minute walk from Quaking Houses, would have little chance of survival. Perhaps a further measure of the project's success, then, is that there has been very little damage. A glass viewing panel in the walkway was broken, but for the most part the wetland flourishes.

Hopkins asked what the world would be, bereft of wilderness. Today we ask what it might be with clean water, fresh air and a restoration of the habitats which ensure bio-diversity. The wetland is part of that future, and not an isolated example. Since the project was begun, another wetland has been created at the Earth Centre near Doncaster, on the site of two abandoned coal mines. This is a national site demonstrating sustainable ecologies, where all the water and sewage on the site passes through reed beds and other organic means of purification. There, too, are living willow sculptures, such as the willow ball by Jim Buchannan, sited in its wetland beside the river Don [see photo, p132]. So, the wetland at Quaking Houses is part of a future which begins to take shape; and the work of three artists, and co-ordination of the project through the Artists' Agency, has contributed to a re-thinking of what we mean by a civilised world. Perhaps it is not a matter of ever-higher buildings, richer corporations and wider markets for things we do not need, but of something much simpler. Long live the weeds and the wilderness yet.

10 *Lessons and conclusions*

What has been learned from the Quaking Houses experience? Some of the issues have been discussed in earlier chapters: the need for a dedicated and independent co-ordinator to draw together the various interest groups in interdisciplinary projects; the problems of communication and overcoming conflicts of interest between the various parties. As partnership working and cross-disciplinary projects become ever more common—arts and science are coming together with increasing frequency in areas such as healthcare as well as environmental projects—it is necessary that participants should be able to communicate effectively. The purposely inexact and conceptual language of the artist is not necessarily comprehensible to the engineer, whose own ideas and language are conditioned by the scientific rigours of defining, testing and measuring; nor might the artist truly understand the engineer. The co-ordinator of such a project as Quaking Houses needs to be multilingual.

In addition, participants in collaborative projects need to have the flexibility and the vision to project themselves into the standpoint of the other participants in a project, as well as to share their own knowledge and beliefs. Seen & Unseen demonstrated how over-anxiety in protecting individual knowledge and areas of responsibility—turning one part of the process into an impenetrable 'mystery'—led to breakdown, and a great hiatus in the project; while the sharing of knowledge, attempted and achieved, while also leading to differences between the participants, contained the seeds of resolution. So Helen Smith schooled herself in the technology of the wetland project, the better to be able to reflect the scientific processes in her artistic input; so the work of the Newcastle University team was motivated by a desire to find a holistic solution to the problems of the Stanley Burn; and so members of the Quaking Houses community were able to bring outsiders into their world and share control over what was happening to it.

The project has created a complex and expanding network of communication, between the three main partners and outwards to a wider world. Most of the information generated has been thought-provoking, much of it challenges the conventional world view as seen even from a relatively poor corner of a rich western nation. Aid, and influence, is 'supposed' to flow from the west to less favoured nations, but the start of the project was in part influenced by the example of villagers in Bolivia; in turn, the influence of the Quaking Houses wetland is capable of being exported to provide low-technology and sustainable solutions to problems of water pollution worldwide. Links have been forged between different disciplines, and work is continuing to present different aspects of the project, with expanding internet material and planned virtual reality/CD ROM, to enable larger and more diverse audiences to learn about the wetland. While the story of Quaking Houses might discourage some from collaborative projects, others will see the huge potentials that exist, learn from Quaking Houses, and design their own collaborative models. Even the authors of this book, who have been only tangentially involved with the project, have been moved by it and as a consequence are able to help by spreading information about it.

And what of the community? The preceding chapters have highlighted the collaboration and the conflicts inherent in the collaborative process, but focusing mainly on the relationship between the successive artists to the project, Artists' Agency, and the university team. The role of the community, of the Quaking Houses Environmental Trust, might be perceived from this as that of a Cinderella, the overlooked partner, while art and science argue their respective corners. Yet if the Quaking Houses story could only have one message, it should surely be this: that people can make a difference. The village of Quaking Houses was expected to die—but its people refused to comply with expectations and instead set about reinvigorating their community and the land in which it lies.

It would be unreasonable to expect that everybody in the village was equally involved: they were not. Not all were interested, or saw the state of the burn as relevant to their lives, or were sympathetic to the involvement of outside bodies in the project. Quaking Houses is not some mythic, idealised, utopian community; like everywhere else, it has different groups, tensions and rivalries. But, while not all of the village was heavily involved in the project, a broad range of people, young and old, did come together around the nucleus of the Environmental Trust. They fought, lobbied and cajoled to get things done, and much has been achieved, partly through lucky breaks but mainly through hard work. Underpinning the entire wetland project has been the ceaseless activity of local people: working for change, finding the right people to provide specialist help and input, monitoring, digging, clearing, planting. If the local community has been the quiet partner in the sometimes fractious debates about the project and its direction, this was not for want of effort, understanding, or belief in its value. And, while not absolving local and central government of the responsibility they bear for Quaking Houses and thousands of other communities, the wetland project indicates what can be achieved when communities empower themselves to cleanse and beautify their land rather than wait for the uncertain promise of benefits from on high.

Bibliography

(Please note that a separate bibliography is attached to Chapter 8, 'The Scientific and Engineering Context of the Quaking Houses Community Wetland'.)

John Barr: 'Durham's Murdered Villages', in *New Society*, 3 April 1969

O.Barrass: *A History of South Moor Cricket Club* (Durham 1993)

Robert Colls: *The Collier's Rant—Song and Culture in the Industrial Village* (London 1977)

Durham County Council: County Development Plan 1951

Durham Federation of Women's Institutes: *The Durham Village Book* (Newbury 1992)

Norman Emery: *The Coalminers of Durham* (Stroud 1992)

John Griffiths: 'Quake Protection', in *Water Bulletin*, 16 January 1998

John W. House: *Industrial Britain: The North East* (Newton Abbot 1969)

Norman McCord: *North East England The Region's Development 1760-1960* (London 1979)

South Moor Local History Group: *South Moor and District in Old Picture Postcards* (European Library, Zaltbommel, The Netherlands 1996)

J. B. Priestley: *English Journey* (London 1934)

James Pritchard: *Leachate Discharge from a Large Body of Contaminated Land: Morrison Busty Spoil Heap, County Durham* (unpublished MSc dissertation, University of Newcastle upon Tyne August 1997)

Shakwi Srour: *Modelling Groundwater Flow through a Body of Contaminated Land: Morrison Busty Spoil Heap, County Durham* (unpublished MSc dissertation, University of Newcastle upon Tyne, August 1998)

David Temple: *The Collieries of Durham* vol 2 (Newcastle upon Tyne n.d.)

Harry Thompson: *Durham Villages* (London 1976)

F. J. Wade: *The Story of Annfield Plain and District* (1966 repr 1986)

F. J. Wade: *The Story of South Moor* (unpub. manuscript 1966)

F. J. Wade: unpublished letter dated 19th June 1969 to Mr Warren on the origins of the name 'Quaking Houses' (Durham City Library)

P. L. Younger (Ed): *Minewater Treatment Using Wetlands: Proceedings of a National CIWEM Conference, 5th September 1997, University of Newcastle upon Tyne* (London 1997)

P. L. Younger: 'The Gavinswelly Wetland', in *Northern Review* vol 2 Winter 1995 (Newcastle 1995)

P. L. Younger, T. P. Curtis, A. Jarvis and R. Pennell: 'Effective Passive Treatment of Aluminium-Rich, Acidic Colliery Spoil Drainage using a Compost Wetland at Quaking Houses, County Durham', in *Journal of the Chartered Institute of Water and Environmental Management*, vol 11 June 1997

Appendix

Scientific data from
Quaking Houses wetland

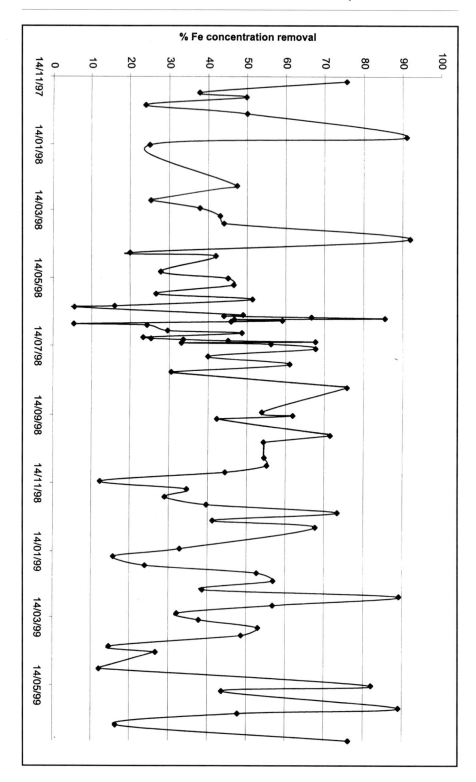

% Fe concentration removal

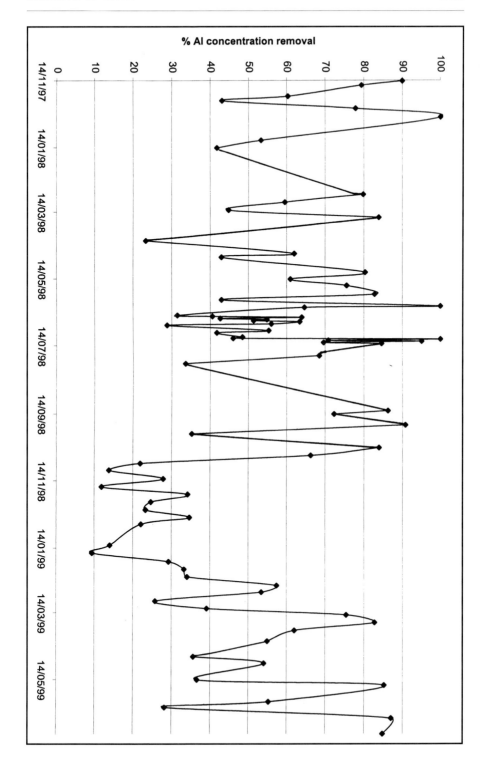

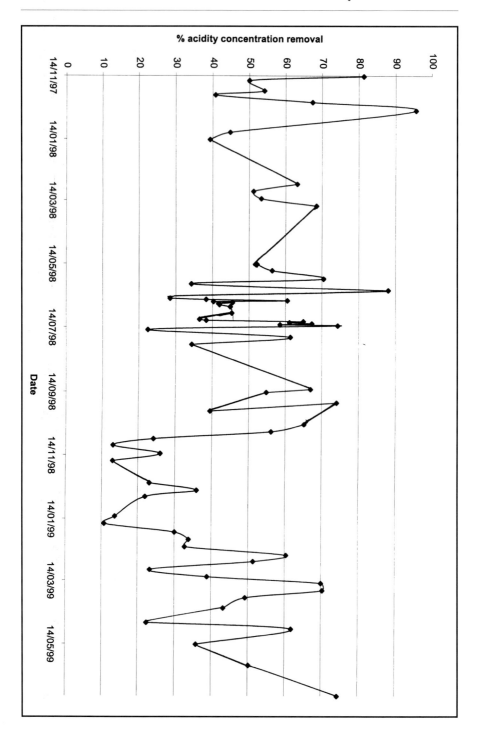